ALLE ZEIT WACH 1842

Azospirillum IV

Genetics · Physiology · Ecology

Proceedings of the
Fourth Bayreuth Azospirillum Workshop

Edited by Walter Klingmüller

With 82 Figures

Springer-Verlag
Berlin Heidelberg New York
London Paris Tokyo

Prof. Dr. WALTER KLINGMÜLLER
Lehrstuhl für Genetik
Universität Bayreuth
Universitätsstraße 30
8580 Bayreuth, FRG

The workshop was sponsored by
BASF Ludwigshafen
Bayerisches Staatsministerium für Ernährung, Landwirtschaft und Forsten
Oldenburgische Landesbank

ISBN-13:978-3-642-73074-0 e-ISBN-13:978-3-642-73072-6
DOI: 10.1007/978-3-642-73072-6

Softcover reprint of the hardcover 1st edition 1988

2131/3130-543210

PREFACE

On June 17th and 18th 1987 the fourth workshop on "Azospirillum: Genetics, Physiology, Ecology" took place at the University of Bayreuth, West Germany, organized by the Genetics department. There were about 80 participants, who came from German research institutions, from other European countries, from Israel, Egypt, India, North Vietnam, and North and South America. The former workshops had also taken place in Bayreuth, in 1981, 1983 and 1985, hence the organizers could draw on the experience obtained at these earlier workshops.

Azospirilla have, during the past 15 years, found an ever increasing scientific interest because, first, these soil bacteria carry the genetic information for binding molecular nitrogen from the air and, second, they live in close vicinity to the roots of grain crops and forage grasses. By exploitation of these two properties, it is hoped to develop inoculation procedures which can be used in nitrogen-deficient soils.

The reports on the first afternoon focussed, as in the former workshops, on genetic approaches. Considerable progress has again been made, via the application of refined gene technological methods, in the understanding of nitrogen fixation of both *Azospirillum brasilense* and *Azospirillum lipoferum*. Physical maps were presented not only for the structural nif-genes but also extending into neighbouring regions and three additional genes, among them one of regulatory type, were identified. Also the understanding of other bacterial functions was improved by the induction and characterization of different types of auxotrophic, growth hormone negative and association defective mutants, in particular via transposon mutagenesis. The latter method is now well developed for application in Azospirilla.

Further metabolic and ecological properties of Azospirilla were broadly discussed, e. g. carbon source and nitrate utilization, bacteriocin production, nitrogenase sensitivity against oxygen, salt tolerance. Ways to measure association quantitatively, with an eye to screening for association defective mutants were suggested, and DNA regions homologous to the nod and hsn genes in Rhizobium were identified, that could be involved in the association process. Electron microscopic investigations, in combination with specific fixation and staining procedures, provided convincing evidence for the presence of Azospirilla upon inoculation not only on the root surface but also in the intercellular spaces of the root cortex, and even inside root cells.

Reports on actual beneficial effects of inoculations were few, the discussions on this point were somewhat reluctant or controversial, in particular for temperate climates. Also, there was no consensus on the importance of N_2 fixation for such effects. Azospirilla have several other properties, which indirectly could cause beneficial effects. However in this regard they do not differ significantly from many other soil bacteria. Hence the idea came up, to extend the perspective to more and different bacterial species instead of focussing on Azospirilla alone in such studies. The isolation of several new strains similar to Azospirilla from different parts of the world was reported. It was also stressed that, due to competing bacteria in the soil and to ecological factors, the effects of inoculations, would always depend markedly on time of inoculation, survival of bacteria in the soil, the bacterial titer close to the roots, and even on the plant variety and its exudates leaking from the roots, not mentioning other things. Hence, the situation is as complex as ever.

However it can be stated that even if Azospirilla cannot yet, in the near future, be developed to become suitable and reliable biofertilizers in agriculture, they will certainly continue to be fascinating model systems for basic research on soil bacteria, on the border line between symbiotic and free-living nitrogen fixers. As such they will play an ever increasing role in scientific research. Their importance for practical applications will clearly be larger than that of Klebsiella or E. coli.

The number of contributions, as well as the number of participants of the workshop documented the need for such workshops and the importance of the subject. The opportunity for an intense exchange of experiences and ideas was exploited not only during the scientific meetings, but also during informal events. Again, many new personal contacts were established. It is hoped that these will pay off in the future. The foreign participants who had come for the first time were impressed by Bayreuth University, its research institutions, and the genetics department. An evening at the summer residence "Schloßhotel Thiergarten" in its scenic setting added positively to the overall impression.

Thanks to the speakers, who had been asked to bring the manuscripts with them ready for print, and thanks to the publisher, it is again possible to bring these proceedings onto the market within three months after the workshop. They will help the participants to recapitulate and better understand the contributions. They should moreover help all other interested people, to judge the present state of Azospirillum work, and to evaluate the prospects of working with this organism in the future.

W. KLINGMÜLLER

CONTENTS

LIST OF PARTICIPANTS

Abd-el-Salam, M.	Cairo	Egypt
Alef, K.	Bayreuth	Germany
Antonelli, N.M.	Bari	Italy
Apfel, Ch.	Bayreuth	Germany
Barbieri, P.	Milano	Italy
Bashan, Y.	Rehovot	Israel
Bazzicalupo, M.	Firenze	Italy
Beyer, D.	Bayreuth	Germany
Beyse, J.	Bayreuth	Germany
Bieleit, G.	Braunschweig	Germany
Brandmüller, E.	Bayreuth	Germany
Christiansen-Weniger,Ch.	Wageningen	Netherlands
De Coninck, K.	Heverlee	Belgium
Del Gallo, M.	Rome	Italy
Dickmanns, E.A.	Bayreuth	Germany
Döhler, K.	Bayreuth	Germany
Drozdowicz, A.	Rio de Janeiro	Brasil
Dums, F.	Bayreuth	Germany
El-Khawas, H.	Cairo	Egypt
Elmerich, C.	Paris	France
Engels, A.	Iserlohn	Germany
Eyers, M.	Leuven	Belgium
Favilli, F.	Firenze	Italy
Fischer, E.	Bayreuth	Germany
Fogher, C.	Piacenza	Italy
Fritzsche, Ch.	Hannover	Germany
Gadkari, D.	Bayreuth	Germany
Hartmann, A.	Bayreuth	Germany
Horemans, S.	Leuven	Belgium
Hurek, T.	Hannover	Germany
Jagnow, G.	Braunschweig	Germany

Kimmel, S.	Hannover	Germany
Klaus, I.	Bayreuth	Germany
Klingmüller, U.	Heidelberg	Germany
Klingmüller, W.	Bayreuth	Germany
Krämer, J.	Bonn	Germany
Krauß, T.	Bayreuth	Germany
Kreutzer, R.	Bayreuth	Germany
Langenbach, T.	Rio de Janeiro	Brasil
Levanony, H.	Rehovot	Israel
Markus, P.	Bonn	Germany
Martin, P.	Hohenheim	Germany
Mummenthey, K.	Bayreuth	Germany
Nayak, D.N.	Orissa	Indien
Neyra, C.A.	New Brunswick	USA
Nguyen ngoc Dung	Hanoi	Vietnam
Niemann, E.G.	Hannover	Germany
Raps, C.	Bayreuth	Germany
Reiß, J.	Bayreuth	Germany
Reinhold, B.	Hannover	Germany
Ruckdäschel, E.	Bayreuth	Germany
Schilf, W.	Bayreuth	Germany
Schirmer, N.	Bayreuth	Germany
Schmidt, A.	Bayreuth	Germany
Schmidt, W.	Hohenheim	Germany
Schmitt, M.	Bayreuth	Germany
Schumann, W.	Bayreuth	Germany
Schwabe, G.	Essen	Germany
Singh, M.	Köln	Germany
Steibl, H.-D.	Bayreuth	Germany
Stanzel, K.	Bayreuth	Germany
Stumpf, F.	Bayreuth	Germany
Tolentino, D.	Novara	Italy
Ueckert, J.	Hannover	Germany
Van Elsas, J. D.	Wageningen	Netherlands
Vanstockem, M.	Leuven	Belgium
Vlassak, K.	Leuven	Belgium
Weihs, V.	Braunschweig	Germany
Werner, C.	Bayreuth	Germany
Zaha, A.	Porto Alegre	Brasil
Zirbel, M.R.	Bayreuth	Germany

ADVANCES IN GENETICS OF AZOSPIRILLUM BRASILENSE Sp7: USE OF TN5 MUTAGENESIS FOR GENE MAPPING AND IDENTIFICATION

M. Galimand, C. Vieille, B. Perroud, I. Onyeocha and C. Elmerich.

Unité de Physiologie Cellulaire, Département des Biotechnologies,
Institut Pasteur,28 rue du Docteur Roux,
75724 Paris Cedex 15, France.

Summary : Tn5 site-directed mutagenesis was used to identify nif genes and to localize plasmid genes from A. brasilense Sp7 (ATCC29145). A DNA region covering 30 kb and containing the nitrogenase structural genes (nifHDK) has been cloned. The presence of nif genes in the 20 kb located next to nifHDK was explored by Tn5 mutagenesis and two new nif loci were identified. Homology between Rhizobium meliloti nodulation genes (nod and hsn) and Azospirillum total DNA was previously reported. A 10 kb EcoRI fragment homologous to hsn was cloned in pUC18, to yield pAB502. Preliminary experiments suggested that the hsn homologous region was located on the 90 Md plasmid contained in strain Sp7. This was confirmed after mutagenesis of pAB502 by Tn5-mob and recombination of the insertions in the 90 Md plasmid. Key words : Azospirillum, nif cluster, Tn5 mutagenesis, plasmid.

Introduction

In many Gram negative bacteria, transmissible plasmids can be utilized to introduce transposon in the host genome (1). In particular, suicide plasmids which can be mobilized but cannot replicate in the host bacteria have been proved to be useful tools to perform random and localized mutagenesis (2,3). In Azospirillum several reports already described conditions of Tn5 mutagenesis in order to identify and to clone genes involved in nitrogen fixation (4-8). We previouly reported experiments of site-directed Tn5 mutagenesis of the nifHDK cluster, encoding nitrogenase stuctural genes in A. brasilense Sp7, using pSUP202 as a suicide vehicle (4). This vector, a derivative of pBR325, contains the site for mobilization (mob region) of incompatibility P type plasmids and cannot replicate in Azopirillum (2). Results were in agreement with the existence of a single operon transcribed in the nifHDK order as in

Azospirillum IV: Genetics, Physiology, Ecology
Edited by W. Klingmüller

Klebsiella pneumoniae (4). We now report the use of Tn5 mutagenesis to identify nif genes in the vicinity of nifHDK and to characterize a DNA region from the 90 Md plasmid of strain Sp7.

Materials and Methods

Strains, plasmids, media and growth conditions:

Strains are listed in Table 1. Plasmids are listed in Table 1 and described in Figure 1 and 2. Media for Azospirillum and Escherichia coli were as previously described (4). Assay for nitrogen fixation in whole cells was performed with overnight cultures grown in the presence of lactate and glutamate as reported (4).

TABLE 1: Strains and plasmids

Strain	Genotype or phenotype	Reference
A. brasilense		
Sp7 (ATCC29145)	wild type	(9)
7030	Sp7 mutant, lack the 115 Md plasmid	(10)
E. coli		
C600	trp, leu, lacY, recA, rif, r^-m^-	(4)
JC5466(Tn5-mob)	trp, his, recA, rpsE, rif,	this work
	contains Tn5-mob (Km^R)	(4)
S17.1	pro, thi, RP4 (Tc::Mu Km::Tn7)	(2)
S17.1::Tn5	contains Tn5 (Km^R)	(4)

Plasmid	Vector	Phenotype	Reference
pRK2013		Km^R Tra^+	(11)
pUC8, pUC18		Amp^R, lacZ	(12)
pSUP202	pBR325	Amp^R, Cm^R, Tc^R, Mob	(2)
pAB3	pSUP202	Amp^R, Tc^R, Mob, AbRI	(4)
pAB10, pAB16	pUC8	Amp^R, Sp7 DNA	(Fig. 1)
pAB14, pAB19	pSUP202	Amp^R, Cm^R, Mob, Sp7 DNA	(Fig. 1)
pAB502	pUC18	Amp^R, 10 kb EcoRI fragment	(Fig. 2)

Tn5 mutagenesis

Mutagenesis of pAB14 and pAB19 was performed in S17.1::Tn5 as described (4), except that transconjugants were selected in strain C600. Plasmids carrying Tn5 were subsequently introduced into strain Sp7, using pRK2013 as mobilizing agent (11). Mutagenesis of pAB502 was performed in JC5466(Tn5-mob) using kanamycin gradient plates (13). Tn5-mob containing plasmids were extracted from the Km^R mutants and introduced in S17.1 by transformation. Azospirillum recombinants were selected on minimal medium containing 20 µg/ml kanamycin.

Molecular biology techniques

Plasmid construction, transformation, DNA isolation, restriction analysis and DNA hybridization employed conventional techniques (14). Plasmid extraction from Azospirillum was performed according to Kado and Liu (15).

Results and discussion

Identification of two nif clusters in the vicinity of nifHDK.

Cloning of DNA adjacent to nifHDK: In Azospirillum Sp7, the nifHDK genes are carried by a 6.7 kb EcoRI fragment (16). This fragment was cloned at the unique EcoRI site of pSUP202 to yield pAB3 (4). A kanamycin cartridge, corresponding to the 2.2 kb XhoI fragment purified from Tn5 and containing a single HindIII site, was inserted at the unique XhoI site of pAB3 (Fig. 1). After recombination in Sp7 genome, the resulting Nif^- strain was used to recover a 10 kb HindIII fragment (pAB10 in Fig. 1), containing the DNA region located downstream from the nifHDK operon (17). From a Tn5 insertion located at the extremity of the newly cloned region, it was possible to recover a 15 kb BamHI fragment (pAB16 in Fig. 1), containing the adjacent region (17).

Tn5 mutagenesis: Site-directed Tn5 mutagenesis was performed on pAB14 and pAB19 which were subcloned from pAB10 and pAB16 into pSUP202. More than 50 different Tn5 insertions were isolated in pAB14 or pAB19 and subsequently recombined in Sp7 genome (17). Two regions leading to Km^R mutants with Nif^- phenotype were identified, as indicated in Figure 1. It was checked by hybridization with a Tn5 probe and with a nif probe that Tn5 has recombined at the right location. Hybridization with K. pneumoniae and Rhizobium ORS571 nifE probes suggested that the newly identified nif region proximal to nifK contained the equivalent of a nifE gene (17,18). The third region might contain DNA homologous to nifMVSU.

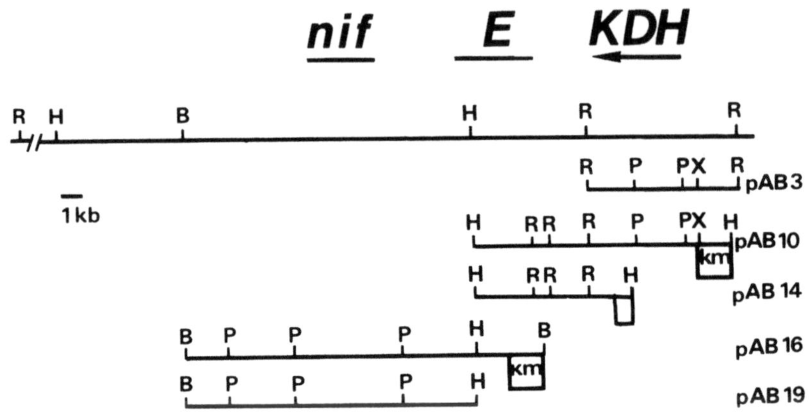

Figure 1: Physical map of the DNA region of A. brasilense Sp7, containing nif DNA. NifHDK and nifE genes were identified by hybridization with K. pneumoniae or Rhizobium ORS571 nif probes. Arrow: direction of transcription. Restriction sites: B: BamHI; H: HindIII; P: PstI; R: EcoRI; X: XhoI. Open square indicate a Tn5 fragment or a kanamycine (Km) cartridge.

In K. pneumoniae, 17 nif genes (nifQBALFMVSUXNEYKDHJ) clustered on a DNA fragment of 22 kb were characterized (see review in Ref. 19). We have cloned the equivalent of nifHDK and possibly nifE from A brasilense Sp7. Regulatory mutants displaying a nifA phenotype have been isolated (20). In addition, hybridization with fix genes from Rhizobium revealed homology to fixA (21). Using random Tn5 mutagenesis Vanstockem et al. (8) identified a new nif locus in Sp7 carried by a 4.2 kb EcoRI fragment. From strain 29710, which is close to Sp7, Singh and Klingmüller (7) have cloned three EcoRI fragments of 15, 16 and 17 kb containing nif loci. Genes from these new nif loci have not been identified as yet. This suggests that the nif genes in A. brasilense are more scattered than in K. pneumoniae and that

Azospirillum possibly contains other nif genes, not found in K. pneumoniae, which remain to be identified.

Homology between Rhizobium nodulation genes and Azospirillum total DNA and cloning of Azospirillum homologous DNA fragments.

We previously reported homology between Rhizobium nodulation genes and Azospirillum total DNA (21). This observation may reflect a common mechanism in the early process of bacterial association to plants. In order to clone the homologous fragmnents from Sp7, gene banks constituted of Azospirillum Sp7 EcoRI or SalI fragments were constructed in pUC18 (22). Colony hybridization was used to identify clones that carried homology to either nod or hsn probes. A recombinant plasmid, pAB502, carried a 10 kb EcoRI fragment and the extent of the homology with the hsn region was limited to a 1.4 kb XhoI-SmaI fragment. Figure 2 shows the physical map of the cloned fragment and the approximate localization of the homology region.

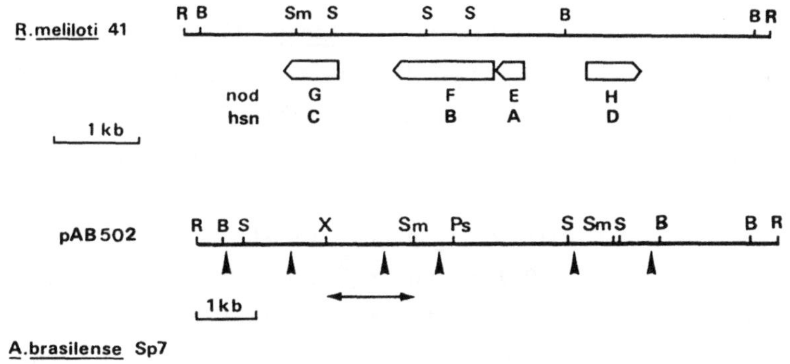

Figure 2: Physical maps of the A. brasilense Sp7 10 kb EcoRI fragment from pAB502 and of R. meliloti hsn region. Genes in the hsn region are designated either nodEFGH or hsnABCD (23). Restriction sites: as in Fig. 1; S: SalI; Sm: SmaI. Double-headed arrow shows the homology region to the hsn probe. Black triangles indicate the location of Tn5-mob insertions.

Use of Tn5 mutagenesis to characterize a DNA fragment fom the 90 Md plasmid of A. brasilense Sp7.

In rhizobia, functions related to symbiotic nitrogen fixation were shown to be plasmid borne (23, see review in Ref. 24). All *Azospirillum* strains, examined so far, contained at least one plasmid (see review in Ref. 25). No phenotypic property has been demonstrated as plasmid-borne, except for the phage All which is present in the lysogens as a plasmid (25). However, it is tempting to speculate that functions related to bacteria-plant association might be present on *Azospirillum* plasmids. To determine the chromosomal or plasmidic localization of the *hsn* homologous region, Southern blot hybridization was performed. *A. brasilense* Sp7 contains 4 plasmids of approximate molecular weight 90, 115 and over 300 Md, which can be visualized in agarose gels (10). Mutant 7030 has lost the 115 Md plasmid (10). Hybridization of plasmid preparation with pAB502 *Eco*RI fragment suggested that the fragment originated from the 90 Md plasmid, p90, of Sp7 (data not shown).

Mutagenesis of pAB502 was performed in *E. coli* JC5466 containing Tn*5*-*mob* inserted in the chromosome. This transposon contains a 3 kb *Bam*HI fragment carrying the *mob* region of RP4. When introduced in pUC18 the resulting plasmid can be transferred in *Azospirillum* and used for random or site directed mutagenesis. Six different insertions were isolated in the 10 kb *Eco*RI fragment (Fig. 2) and recombined in Sp7 genome. Plasmid was extracted from the recombinants and after transfer to nitrocellulose filter hybridized with Tn*5*. Figure 3 shows that the p90 from the km^R recombinants hybridized with a Tn*5* probe. In addition, an increase in the molecular weight of the p90 from the Km^R recombinants was observed. This strongly favors the plasmidic origin of the cloned region.

It is too early to conclude on the significance of the homology detected between *nod* genes and *Azospirillum* DNA, since we do not know yet if the corresponding regions are functional in *Azospirillum*. However, it is interesting to note that one of the homologous fragments originated from a plasmid. This fragment can now be used as a probe to study relationship between plasmids from *Azospirillum* strains.

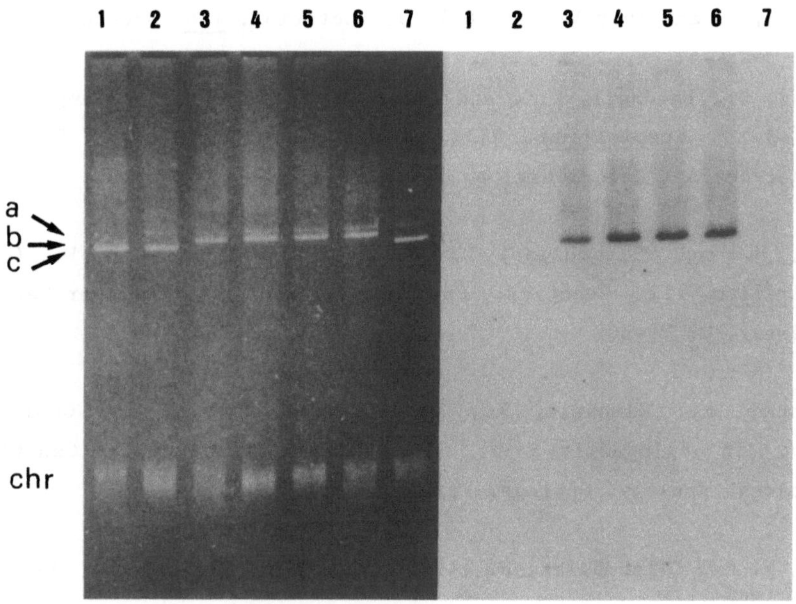

Figure 3: Localization of the 10 kb EcoRI fragment from pAB502 in the p90. Left: agarose electrophoresis of plasmid extracts; right: corresponding autoradiogram after hybridization with Tn5. 1: strain 7030 (devoid of p115); 2: Sp7 spontaneous Km^R mutant; 3-6: Sp7 Km^R (Tn5-mob) recombinants; 7: Sp7; a: p115; b: p90-Tn5-mob; c: p90; chr: chromosomal DNA. Megaplasmids of over 300 Md are not visible in this experiment.

Acknowledgements

C.V and BP were recipient of a predoctoral fellowship from the Ministère de l'Industrie et de la Recherche, I.O was the recipient of a predoctoral fellowship from the French Government. This work was supported by funds from the University Paris VII, and by a research contract from Elf Bio-Recherche, Entreprise Minière et Chimique, Rhône-Poulenc Recherche, and CDF Chimie.

References

1. Ruvkun, G.B. and Ausubel F.M. (1981). Nature 289, 85-88.

2. Simon, R., Priefer,U. and Pühler, A. (1983). Bio/Technology 1, 784-789.

3. Selvaraj, G. and Iver V.N. (1983). J. Bacteriol. 158, 580-589.

4. Perroud, B., Bandhari, S.K. and Elmerich, C. (1985). In Klingmüller, W. (ed.) *Azospirillum* III: Genetics, Physiology, Ecology. Springer-Verlag, Heidelberg, p. 10-19.

5. Singh, M. and Klingmüller, W. (1985). In Klingmüller, W. (ed.) *Azospirillum* III: Genetics, Physiology, Ecology. Springer-Verlag, Heidelberg, p. 29-30.

6. Vanstockem M., Michiels, K., Vanderleyden, J. and Van Gool, A. (1985). In Klingmüller, W. (ed.) *Azospirillum* III: Genetics, Physiology, Ecology. Springer-Verlag, Heidelberg, p. 74-84.

7. Singh, M. and Klingmüller, W. (1986). Mol. Gen. Genet. 202, 136-142.

8. Vanstockem, M., Michiels, K., Vanderleyden, J. and Van Gool, A. (1987). Appl. Environm. Microbiol. 43, 410-415.

9. Tarrand, J.J., Krieg, N.R. and Döbereiner, J. (1978). Can. J. Microbiol. 24, 967-980.

10. Franche, C. and Elmerich, C. (1981). Ann. Microbiol. (Inst. Pasteur) 132A, 3-19.

11. Ditta, G., Stanfield, S., Corbin, D. and Helinski, D. (1980). Proc. Natl. Acad. Sci. USA 77, 7347-7351.

12. Yanish-Perron, C., Vieira, J. and Messing, J. (1985). Gene 33, 103-119.

13. Simon, R. (1984). Mol. Gen. Genet. 196, 413-420.

14. Maniatis, T., Fritsch E.F., and Sambrook J. (1982). Molecular cloning, Cold Spring Harbor Laboratory.

15. Kado, C.I. and Liu, S.T. (1981). J. Bacteriol. 145, 1365-1373.

16. Quiviger, B., Franche, C., Lutfalla, G., Rice, D., Haselkorn, R. and Elmerich, C. (1982). Biochimie 64, 495-502.

17. Elmerich, C., Bozouklian, H., Vieille, C., Fogher, C., Perroud, B., Perrin, A. and Vanderleyden, J. (1987). Phil. Trans. R. Soc. Lond. B, in press.

18. Elmerich, C., Fogher C., Bozouklian H., Perroud B., and Dusha I. (1985). In Evans, H. J., Bottomley, P.J. and Newton, W.E. (eds.) Nitrogen fixation research progress. Martinus Nijhoff Publishers, Dordrecht p. 477-483.

19. Elmerich, C. (1984). Bio/Technology 2, 967-978.

20. Pedrosa, F. and Yates, M.G. (1984). FEMS Microbiol. Lett. 23, 95-101.

21. Fogher, C., Dusha, I., Barbot, P. and Elmerich, C.(1985). FEMS Microbiol. Lett. 30, 245-249.

22. Vanderleyden, J., Vieille, C., Michiels, Matassi, G., Van Gool, A. and Elmerich, C. (1986). In Lugtenberg, B. (ed.) Recognition in microbe-plant symbiotic and pathogenic interactions. Nato ASI Series Vol H 4. Springer Verlag, Berlin, Heidelberg, p. 215-218.

23. Horvath, B., Kondorosi, E., John, M., Schmidt, J., Török, I., Györgypal, Z., Barabas, I., Wieneke, U., Schell, J., and Kondorosi, A. (1986). Cell 46, 335-343.

24. Prakash, R. K. and Atherly, A. (1986). Int. Rev. Cytol. 104, 1-24.

25. Elmerich C. (1986) In Pühler, A. and Broughton, W.J. (eds.), Nitrogen Fixation Molecular Biology Vol. IV. Clarendon Press, Oxford, p. 106-126.

CLONING OF AZOSPIRILLUM BRASILENSE SEQUENCES PROMOTING TRANSCRIPTION IN ESCHERICHIA COLI

Concetta Schipani, Marco Bazzicalupo, Enzo Gallori, Ettore Luzi, Luca Turbanti and Renato Fani

Dipartimento di Biologia Animale e Genetica, Università di Firenze, Via Romana 17, 50125 Firenze, Italy

Summary. Some genes of *A.brasilense* can be expressed in *E.coli*. We take advantage of this property for the isolation of *A.brasilense* promoter sequences. DNA fragment of *A.brasilense* SP6 were cloned into the BamHI site of plasmid pKK232-8 upstream the chloramphenicol resistance (CAT) gene which lacks its own promoter. Some of the *Azospirillum* DNA sequences selected for the ability to promote CAT gene transcription have been further characterized either for the level of resistance to chloramphenicol and for sequence analysis.
Key words: cloning, transcription promoters

Introduction

In bacteria gene expression is controlled, in part, at the level of RNA transcription; transcriptional regulation is achievede by modulating the efficiency with which the RNA polymerase can recognize and interact with particular sites, promoters, from which the polymerase initiates transcription. Transcription promoters have been mainly and extensively studied in *E.coli*; up to now the sequence of more than one hundred Enterobacteriaceae promoters has been determined and their comparison has permitted the identification of two hexanucleotidic regions of homology, located at -35 and -10 nucleotides upstream the startpoint (1).

In some instances genes that are under the control of the same regulatory protein(s) share the same "consensus sequence", even if the-

Azospirillum IV: Genetics, Physiology, Ecology
Edited by W. Klingmüller

se genes belong to different bacterial species. That is the case of the genes involved in the nitrogen metabolism controlled at a transcriptional level by the ntr system (2). These genes are transcribed from (activable) promoters that lack the canonical -10 and - 35 sequence of E.coli promoters, but they still share consensus sequences in two regions located -24 and -12 upstream the startpoint. These regions are common to ntr controlled genes of several different bacterial species, i.e. E.coli, Klebsiella pneumoniae, Rhizobium (3).

In the nitrogen fixing bacterium Azospirillum brasilense the control of transcription is still poorly understood although many progress in the molecular genetic of this bacterium have been recently realized (4) and regulation of gene expression is known to occurr at least for enzymes involved in nitrogen metabolism (5, 6).

In a previous work we reported (7) that some Azospirillum genes can be expressed in E.coli indicating that, at least in some cases, the signals for RNA polymerase could be similar in the two organisms. For this reason we used E.coli cloning vector in order to identify A.brasilense sequences able to promote transcription.

RESULTS AND DISCUSSION

Cloning of A.brasilense DNA sequences recognized by E.coli RNA-polymerase. We used an E.coli system for the identification of Azospirillum DNA sequences able to promote transcription. For the isola - tion of the"promoters" we used the plasmid pKK232-8 (8) as cloning vector (Fig. 1); it carries the genes coding for the ampicillin and chloramphenicol resistances. The latter gene, coding for a chloramphenicol acetyl transferase (CAT) activity, is not transcribed be -

cause of the lack of the RNA-polymerase recognition sites (-10 and -35 regions). These regions have been replaced by a multiple cloning site which allows the insertion of exogenous DNA sequences. The CAT gene can be expressed if a DNA sequence recognized by RNA-polymerase has been inserted in the upstream polylinker, allowing the cell to grow in the presence of the antibiotic. Moreover the level of resistance to chloramphenicol is directly correlated to the CAT expression and consequently to the transcription promoting strngth of the cloned sequence.

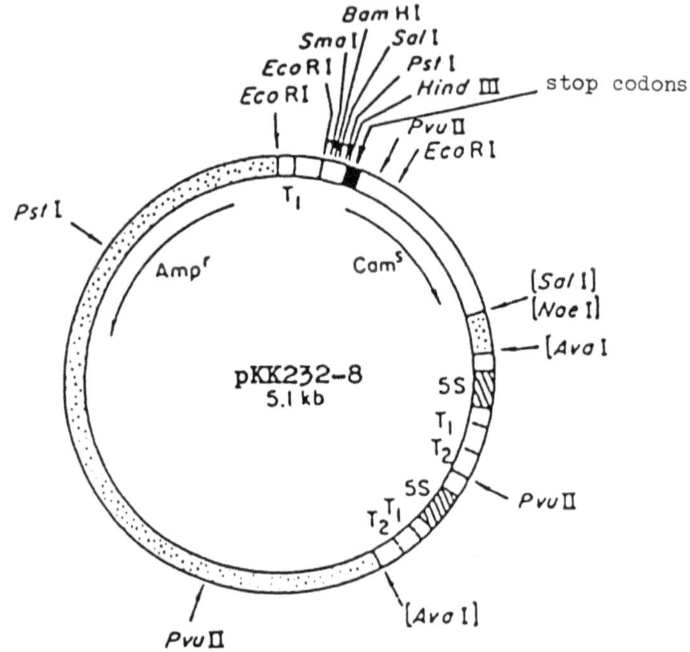

Figure 1. Restriction map of plasmid pKK232-8

A.brasilense total DNA from strain SP6 was digested with restriction endonuclease Sau3A and fragments in the range of 100-400 bp were ligated to plasmid pKK232-8 digested with BamHI. The ligation mixture was used to transform E.coli strain DH1 recA Cm^s made competent by the method of Hanahan (9), selecting on LM medium (9) with ampicillin(100 μg/ml). Transformants were then streacked on LM containing ampicillin (100 μg/ml) and choramphenicol (10 μg/ml), an antibiotic concentration which completely inhibits the growth of the parental strain. Out of more three thousands transformants, we obtained about one hundred clones in which the pKK232-8 CAT gene was transcribed from Azospirillum DNA insert, thus indicating that A.brasilense sequences could promote transcription by E.coli RNA-polymerase.

The strength as "transcription promoters" of the cloned sequence. In order to determine the level of resistance to chloramphenicol, each recombinant clone was streacked on LM with different concentration of the antibiotic. As reported in table 1 most clones were resistant to low concentrations of chloramphenicol (10-15 μg/ml); on the other hand some clones showed an higher level of resistance (over 100 μg/ml).This result could be probably ascribed, in most cases, to the inefficient recognition of A.brasilense insert by RNA-polymerase of E.coli, thus indicating a possible different organization of A.brasilense putative promoters in respect to the E.coli ones.

Some of the recombinant clones were tested for the chloramphenicol acetyl transferase activity as follows: 5 ml from a growing culture were collected and resuspended in 1.5 ml 0.25 M TRIS pH 7.8. Cells were disrupted by sonication and the crude extract centrifuged in an eppendorf tube for 5' at 4°C to spin down debris. The supernatant was used to test the CAT activity in the presence of acetyl-CoA and ^{14}C -chloramphenicol as described (10). All the clones resistant in vivo to the antibiotic also showed detectable CAT activity. The autoradiography reported in figure 2 shows the results of one of such experiments which clearly indicated the strong difference between the parental strain , DH1 (pKK232-8) and two of the recombinant clones DH1 (pAF151) and DH1 (pAF280) which were resistant to different concentrations of chloram -

Table 1. Resistance to chloramphenicol of DH1 recombinant clones

	Growth on LM medium with Cm (µg/ml)								
	10	20	30	40	50	60	70	80	90
N. of clones	50	19	12	9	6	4	2	1	1

phenicol, 70 and 40 µg/ml, respectively.

The A.brasilense DNA fragments of some recombinant plasmids were cut off and inserted in both the orientations in the sequencing vector pEMBL 19 (11). The nucleotide sequence of two of them, pAF151(216 bp) and pAF280 (250 bp), was determined by the Sanger method. The nucleotide sequence, not reported, were computer analized and homologies with E.coli promoters sought. In both cases strong homologies with the canonical -35 and -10 regions were not found. These results could suggest that the structural and/or functional organization of Azospirillum sequences promoting transcription could be different from E.coli usual promoters. We are now trying to insert the CAT gene, together with the putative Azospirillum promoter into a shuttle vector able to replicate both in E.coli and Azospirillum. This system could permit to study the sequences in their original organism; moreover it

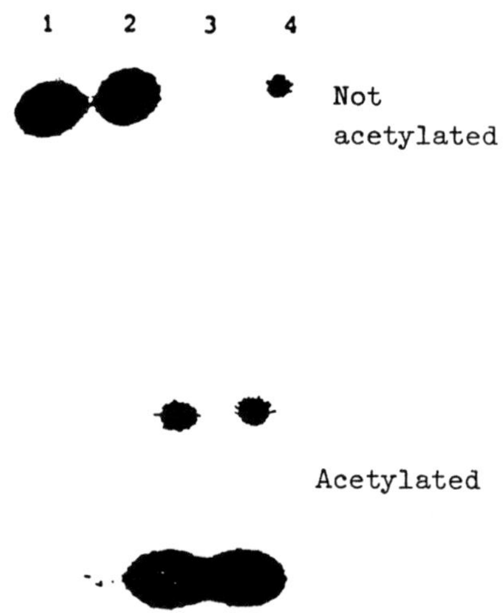

Figure 2. Autoradiography of a TCL plate of CAT assay on extract of strains DH1 (1), DH1 (pKK232-8) (2), DH1 (pAF 151) (3) and DH1 (pAF280)(4).

could be possible to compare the efficiency of the promoting activity in these two organisms.

References

1. Rosenberg,M. and Court D. (1979). Ann.Rev.Genet. 13; 319-353
2. Gussing, G.N., Ronson, C.W. and Ausubel F.M. (1986). Ann.Rev.Genet. 20; 567-592
3. Dixon R.A. (1984). J.Gen.Microbiol. 130; 2745-2755
4. Klingmuller W. (ed) (1985). Azospirillum III: Genetics, Physiology, Ecology. Springer-Verlag, Heidelberg.
5. Bani,D., Barberio, C., Bazzicalupo, M., Favilli, F., Gallori,E. and Polsinelli M. (1980). J.Gen.Microbiol. 119; 239-244
6. Gauthier, D. and Elmerich C. (1977). FEMS Microbiol.Lett. 2;101-104
7. Bazzicalupo, M., Fani, R., Gallori, E., Turbanti, L. and Polsinelli M. (1987). Mol. Gen. Genet. 206; 76 - 80
8. Brosius J. (1984). Gene 27; 151- 170
9. Hanahan D. (1983). J.Mol.Biol. 145; 713-734
10. Shaw W.V. (1975). Methods in Enzymology, Academic Press, N.Y. 43; 740-746
11. Dente, L., Sollazzo, M., Baldari, C., Cesareni, G. and Cortese R. (1985). InGlover D.M. ed. DNA cloning: a practical approach, IRL Press, Oxford-Washington, vol I; 101-107

CHARACTERIZATION OF DNA SEGMENTS ADJACENT TO THE NIFHDK GENES OF AZOSPIRILLUM BRASILENSE Sp7 BY Tn5 SITE-DIRECTED MUTAGENESIS

Elza F. de Araujo, Arnaldo Zaha, Irene S. Schrank and Diógenes S. Santos
Centro de Biotecnologia, Departamento de Genética, Universidade Federal do Rio Grande do Sul, RS, Brasil

Summary: Three recombinants containing DNA inserts of 13 kb (λAb5), 15 kb (λAb2c) and 16 kb (λAb4) were isolated from a gene library of *A. brasilense* Sp7. These DNA inserts are overlapping DNA segments covering approximately 40 kb of the *A. brasilense* genome in the region of the *nifHDK* genes. The λAb2c recombinant containing the *nifHDK* genes localized in a 6.5 kb *Eco*RI DNA fragment was characterized. The *Eco*RI DNA fragments of this recombinant (6.5, 4.15, 2.4, 1.2 and 0.8 kb) were subcloned into the plasmid pACYC184 and after Tn5 mutagenesis, recloned into the plasmid pRK290 and transferred to *A. brasilense*. Five Nif^- mutants were obtained, one with Tn5 inserted in the *nifHDK* genes and four mapping approximately 4 to 5 kb downstream from the *nifHDK* genes.
Key words: *nif* genes, site-directed mutagenesis, gene library, EMBL4 vector

Introduction

In the last years several molecular genetic studies have been done to understand the basic aspects of nitrogen fixation in *Azospirillum*. The nitrogenase structural genes of *A. brasilense* and *A. lipoferum* have been cloned and characterized (1, 2, 3).

Recently we have constructed a gene library of *A. brasilense* Sp7 using the λvector EMBL4. From this gene library we have isolated a recombinant carrying a 15 kb DNA insert containing the nitrogenase structural genes localized in a 6.5 kb *Eco*RI DNA fragment (4). The construction of the gene library using *A. brasilense* DNA partially digested with *Sau*3A, allowed

Azospirillum IV: Genetics, Physiology, Ecology
Edited by W. Klingmüller

us to isolated overlapping DNA segments of a region of the genome containing the nitrogenase structural genes.

In this report we describe the isolation of two new recombinants containing sequences overlapping the ends of the 15 kb DNA insert, and the use of Tn5 site-directed mutagenesis to detect some nif functions in the DNA segments flanking the nifHDK genes.

Material and Methods

1. Bacterial strains, plasmids and bacteriophages: The bacterial strains, plasmids and bacteriophages used in this study are listed in table 1.

Table 1. Bacterial strains, plasmid and bacteriophages

	Relevant characteristics	Reference
Bacteria		
E. coli HB101	pro,leu,thi,lacY, Str^R,r^-m^-,recA	(10)
E. coli Q358	r^-m^-,supE,TonA	(11)
E. coli LE392	r^-m^-,supE,supF,lacY,met	(12)
A. brasilense Sp7	Ap^R	(13)
Plasmids		
pKY2700	Ap^R, ColE1	(14)
pACYC184	Tc^R, Cm^R	(15)
pSA30	Tc^R, nifHDK genes	(16)
pCRA37	Tc^R, hispGDnifQBALFMVSU	(16)
pRK290	Tc^R, IncP, Tra^-	(17)
pRK2013	Km^R, ColE1, Tra^+	(18)
pPH1	IncP, Gm^R, Spc^R, Tra^-	(19)
Bacteriophages		
λ::Tn5	b221rex::Tn5 cI857,Oam29,Pam80	(20)
λAb2c	λEMBL4 - A. brasilense nifHDK	(4)
λAb4	λEMBL4 - A. brasilense DNA	this work
λAb5	λEMBL4 - A. brasilense DNA	this work

2. Media and culture conditions: A. brasilense and E. coli strains were grown in LB (10g tryptone, 5g yeast extract and 5g NaCl). In the conjugation experiments and nitrogenase activity measurements, the A. brasilense strains were grown in NFb (5). The following antibiotics were added to the media when required: Ampicillin (Ap) 40 μg/ml, Tetracycline (Tc) 10 μg/ml, Kanamycin (Km) 30μg/ml, and Gentamicin (Gm) 50 μg/ml.

3. Plaque hybridization: The screening of the A. brasilense gene library was carried out as described by Benton and Davis (6), using the 1.2 kb and 4.15 kb EcoRI DNA fragments of λAb2c (4) as probes.

4. DNA isolation, restriction endonuclease digestion and hybridization: A. brasilense DNA was prepared by the method of Marmur (7). Plasmid DNA was prepared by the cleared lysate method using Triton X-100. Bacteriophages were purified by centrifugation on a CsCl step gradient and the DNA isolated as described by Maniatis et al. (8). DNA was digested with restriction endonucleases according to the instructions of the suppliers. The DNAfragments were separated by eletrophoresis on agarose gels and then transferred to nitrocellulose or nylon membranes (9). Hybridizations were performed as described (8) using ^{32}P-labelled probes.

5. Tn5 site directed-mutagenesis (21): The EcoRI DNA fragments from λAb2c and λAb4 were subcloned in the EcoRI site of pACYC184. The recombinant plasmids in E. coli HB101 were mutagenized with λ::Tn5 as described by Bruijn and Lupski (12). The Tn5 insertions were located by restriction endonuclease analysis. The EcoRI DNA fragments containing Tn5 insertions were recloned in pRK290 and introduced in E. coli HB101 by transformation. The plasmids pRK290-Ab DNA::Tn5 were introduced into A. brasilense Sp7 by a triparental cross using the plasmid pRK2013 to mobilize the pRK290 derivatives. The transconjugants were selected in NFb medium containing Ap, Km and Tc. To select A. brasilense in which a recombination event ocurred transfer-

ring the Ab DNA::Tn5 to the genome, the A. brasilense strains were conjugated with E. coli HB101 containing the plasmid pPH1 which is incompatible with pRK290. The A. brasilense transconjugants were selected in NFb medium containing Ap, Km and Gm. DNA was isolated from the transconjugants and analyzed by Southern blot to confirm the recombination of Ab DNA::Tn5 with the homologous region in the genome.

6. Assay for nitrogenase activity: The acetylene reduction assays on whole cells derepressed for nitrogenase were performed as follows. A. brasilense strains were grown in LB for 24 h at 37° C with shaking. 30 µl of this culture were inoculated in 12 ml flasks containing 5 ml of semi-solid NFb medium (5) and incubated overnight at 35° C. Then 1 ml of acetylene was added to each flask, incubated for 2h at 30° C and then the ethylene production measured.

7. Complementation of A. brasilense Nif⁻ mutants: The pRK290 derivatives containing the A. brasilense EcoRI DNA fragments corresponding to the fragments which have been mutagenized with Tn5, were transferred to A. brasilense NIf⁻ mutants by a triparental conjugation and the transconjugants were selected in NFb containing Ap, Km and Tc. To study a possible complementation of these mutants with K. pneumoniae nif genes, we have transferred the pRK290 recombinant plasmids containing the 8 kb fragment (nifLABQ), 6.2 kb fragment (nifHDK) and 5 kb fragment (nifUVSM) (16) to the A. brasilense mutants. The resulting transconjugants were tested for nitrogenase activity by the acetylene reduction assay.

Results

1. Isolation of recombinants containing A. brasilense DNA adjacent to the 15 kb insert of λAb2c

The 1.2 kb and 4.15 kb EcoRI DNA fragments corresponding to the ends of the 15 kb DNA insert from λAb2c were used as probes to isolate recombinants from the A. brasilense gene

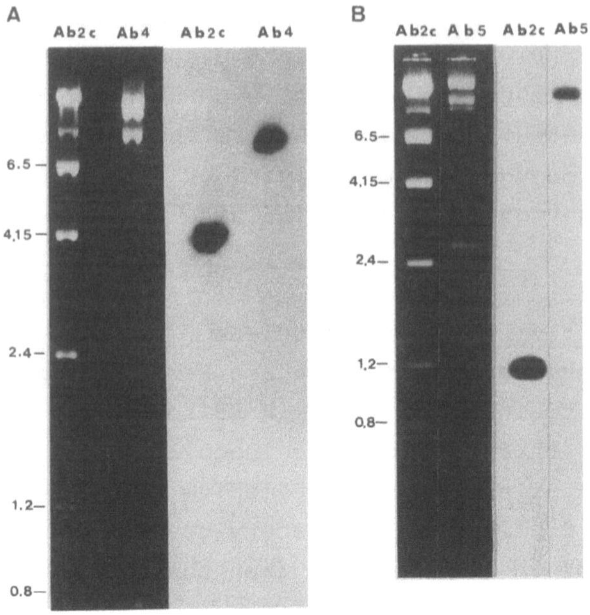

Fig.1. Analysis of recombinants containing sequences flanking the λAb2c DNA insert.
A. λAb4 DNA cleaved with EcoRI and hybridized to the 4.15 kb EcoRI DNA fragment from λAb2c.
B. λAb5 DNA cleaved with EcoRI and hybridized to the 1.2 kb EcoRI DNA fragment from λAb2c.

library. Two recombinants (λAb4 and λAb5) were isolated containing DNA inserts of 16 kb and 13 kb respectively (Fig. 1).

2. Characterization of A. brasilense strains containing Tn5 insertions in the nifHDK and adjacent DNA regions

Fifty-four insertions were obtained in the nifHDK genes and adjacent regions of A. brasilense genome. Figure 2 shows the λAb2c and λAb4 maps with the positions of 54 Tn5 insertions. The Nif phenotypes of some Tn5 insertions were determined. From four different Tn5 insertions in the 6.5 kb EcoRI DNA fragment containing the nifHDK genes, only one (insertion 18) gave a Nif^- phenotype. This insertion was located in the nifD gene. Only one insertion (81) was analyzed in the 2.4 kb EcoRI DNA fragment located on the right of the 6.5 kb EcoRI DNA fragment. This insertion did not give a Nif^- phenotype. Four Tn5 insertions (43, 46,20 and 37) located in the 4.15 kb EcoRI fragment of λAb2c gave a Nif^- phenotype. Insertion 20 was obtained using the vector pACYC184. In this case the 4.15 kb EcoRI fragment mutagenized with Tn5 was introduced into A. brasilense using pRK2013 as the mobilizing plasmid. The phenotype of the different Tn5 insertions analyzed are indicated in Figure 2.

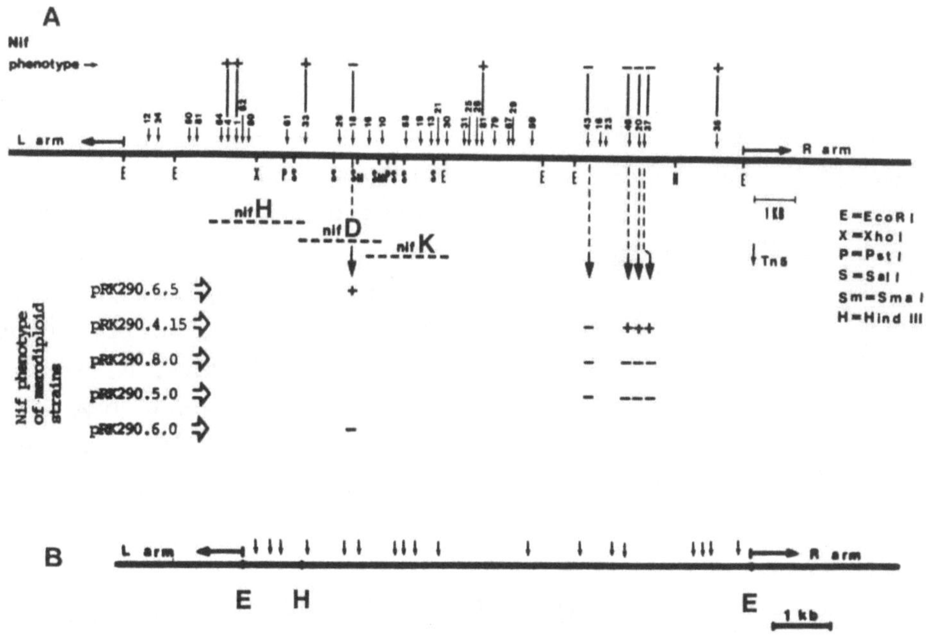

Fig. 2. Physical maps of λAb2c (A) and λAb4 (B) showing Tn5 insertions and the Nif phenotype and complementation analysis of some mutants. The HindIII site in the λAb2c insert corresponds to the HindIII site of the λAb4 insert.

3. Southern blot analysis of A. brasilense transconjugants

The DNA of some A. brasilense strains containing Tn5 insertions were analyzed by the Southern technique to verify the recombination event with the homologous region.

The analysis of DNA from six transconjugants is shown in the autoradiogram of Figure 3. Four A. brasilense transconjugants obtained by the introduction of the plasmid pKR290. 6.5::Tn5 which contains the 6.5 kb EcoRI fragment mutagenized with Tn5 and two transconjugants obtained by the introduction of the plasmid pKR290.2.4::Tn5 containing the 2.4 kb EcoRI fragment mutagenized with Tn5 were analyzed. The results show that five transconjugants containing Tn5 insertions were derived from a recombination event and that one was clearly due to the transposition of Tn5 (Fig. 3B, lane b). We are currently analyzing the insertions derived from the 4.15 kb DNA::Tn5.

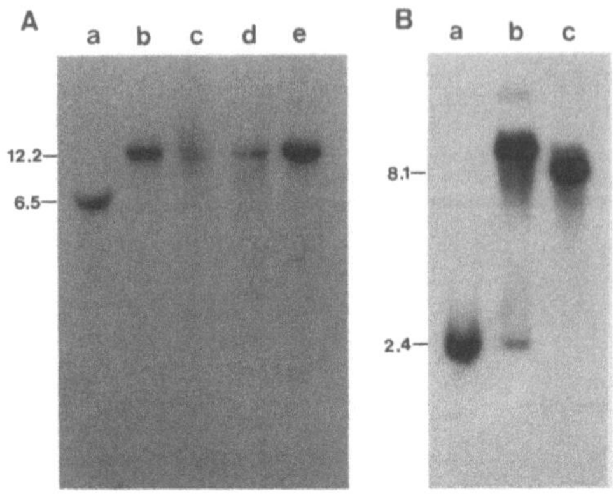

Fig. 3. Southern blot hybridization of some A. brasilense transconjugants containing Tn5. A. Analysis of EcoRI cleaved DNA from transconjugants obtained after introduction of pRK290.6.5::Tn5. a. Sp7, b. Sp7/1, c. Sp7/4, d. Sp7/10, e. Sp7/18. Probe: fragment 6.5::Tn5. B. Analysis of EcoRI cleaved DNA from transconjugants obtained after introduction of pRK290.2.4::Tn5. a. Sp7, b. Sp7/79, c. Sp7/81. Probe: fragment 2.4::Tn5.

4. Complementation analysis of the A. brasilense Nif⁻ mutants

The A. brasilense mutant containing insertion 18 in the nifD gene was complemented by pRK290.6.5 which contains the 6.5 kb EcoRI DNA fragment cloned in the EcoRI site of pRK290. The same strain was not complemented by the plasmid pRK290.6 containing the K. pneumoniae nifHDK genes.

The plasmid pRK290.4.15 carrying the 4.15 kb DNA fragment was introduced in A. brasilense containing insertions 43, 46,20 and 37. After introduction of this plasmid the strains containing insertions 46, 20 and 37 showed a Nif^+ phenotype. The plasmids pRK290.8 and pRK290.5 containing K. pneumoniae nif genes did not complement the Nif^- mutants containing insertions 43, 46 and 37. The Nif^- mutant containing insertion 20 was not tested. The results are summarized in Figure 2.

Discussion

The nif genes have been studied in several nitrogen-fixing microorganisms. The organization of the nif genes in K. pneumoniae are well known. There are 17 nif genes organized in 7 or 8 operons in a 22 kb DNA fragment. Based on the conservation of the nitrogenase genes in different nitrogen-fixing bacteria, it was possible to clone the nifHDK genes from several bacteria using the K pneumoniae nif genes as probes.

In addition to the nifHDK genes, other nif genes have been mapped in some bacteria as for example Azotobacter chroococum (23), A. vinelandii (24), Rhodopseudomonas capsulata (25) and Rhizobium meliloti (26). In these nitrogen-fixing bacteria at least some nif genes are located near the nifHDK genes.

In order to study the organization of the nif genes in A. brasilense we have constructed a gene library using the λvector EMBL4. From this library we have isolated a recombinant (λAb2c) with a 15 kb DNA insert containing the nifHDK genes (4). To characterize a larger region of A. brasilense genome containing the nitrogenase genes we have used the 1.2 kb and 4.15 kb DNA segments flanking the 15 kb DNA insert from λAb2c as probes to isolate new recombinants from the gene library. Two recombinants containing DNA inserts of 16 kb (λAb4) and 13 kb (λAb5) were isolated. These three DNA inserts cover approximately 40 kb of A. brasilense genome in the nifHDK genes region.

In this study we have used the Tn5 site-directed mutagenesis to analyze the possible presence of other nif genes in the DNA adjacent to the nifHDK genes. This method has been used succesfully in the study of nif genes in Rhizobium meliloti (27), A. vinelandii (24), R. capsulata (25) and A. brasilense (2). Using this method we have obtained four Nif$^-$ mutants mapping about 4 to 5 kb downstream to the nifHDK genes. Three of the mutants could be complemented by pRK290.4.15 which contains the 4.15 kb EcoRI DNA fragment from λAb2c, indicating that the Nif$^-$ phenotype was due to a recombination event with the Tn5 mutagenized fragment. These mutants were not complemented by pRK290.8 or pRK290.5 which contain K. pneumoniae nif genes. This result indicates that some nif functions in A. brasilense cannot be substituted for K. pneumoniae genes or that some K. pneumoniae genes are not expressed in A. brasilense. Another possibility would be that the endonuclease used for the subcloning in pRK290 interrupted some fo the K. pneumoniae nif genes.

One of the Nif$^-$ mutant obtained (insertion 18) was mapped in the nifD gene. This mutant was complemented by pRK290.6.5 which contains the A. brasilense nifHDK genes but not by pRK290.6 which contains the K. pneumoniae nifHDK genes. Fahsold et al. (3) have shown that the nifHDK genes from A. lipoferum restored the nitrogenase activity of K. pneumoniae Nif^- mutants only up to 2% of wild type activity and that no complementation was observed with the A. brasilense nifHDK genes.

We are currently characterizing the Tn5 insertions in other sites of the 40 kb DNA segment of A. brasilense containing the nifHDK genes.

Acknowledgements

We are grateful to Dr. J. C. Maia for providing α-^{32}P-dATP and to Dr. F.M. Ausubel for providing bacterial strains. This work was supported by FINEP, PADCT, FAPERGS and CNPq. E.F.A is a CAPES predoctoral fellow.

References

1. Quiviger, B., Franche, C., Lutfalla, G., Rice, D., Haselkorn, R. and Elmerich, C. (1982). Biochimie 64, 495-502
2. Perroud, B., Bandhari, S.K. and Elmerich, C. (1985). In: Klingmüller, W. (ed). Azospirillum III: Genetics, Physiology, Ecology. Springer-Verlag, Heidelberg, p. 10-19
3. Fahsold, R., Singh, M. and Klingmüller W. (1985). In: Klingmüller, W. (ed). Azospirillum III: Genetics, Physiology, Ecology. Springer-Verlag, Heidelberg, p 30-40.
4. Schrank, I.S., Zaha, A., de Araujo, E.F. and Santos, D.S. (1987). Braz. J. Med. Biol. Res. (in press).
5. Döbereiner, J., Mariel, I.E. and Nery, M. (1976). Can.J.Microbiol. 22, 1464-1473.
6. Benton, W.D. and Davis, R.W. (1977). Science 196, 180-183.
7. Marmur, J. (1961). J. Mol. Biol. 3, 208-218.
8. Maniats, T., Fritsch, E. F. and Sambrook, J. (1982). Molecular Cloning. Cold Spring Harbor Laboratory.
9. Southern, E.M. (1975). J. Mol. Biol. 98, 503-517.
10. Boyer, H.W. and Rouland-Dussoix, D. (1969). J. Mol.Biol. 41, 459-472.

11. Karn, J., Brener, S., Barnett, L. and Cesareni, G. (1980). Proc. Natl. Acad. Sci. USA 77, 5172-5176.

12. Bruijn, F.J. and Lupski, J.R. (1984). Gene 27, 131-149.

13. Tarrand, J.J., Krieg, N.R. and Döbereiner, J. (1978). Can. J. Microbiol. 24, 967-978.

14. Ozaki, L.S., Kimura, A., Shimada, K. and Takagi, Y.(1982). J. Biochem. (Japan) 91, 1155-1162.

15. Chang, A.C.Y. and Cohen, S.N. (1978). J. Bacteriol. 134, 1141-1156.

16. Cannon, F.C., Riedel, G.E. and Ausubel, F.M. (1979). Molec. Gen. Genet. 174, 59-66.

17. Ditta, G., Stanfield, S., Corbin, D. and Helinski, D.R. (1980). Proc. Natl. Acad. USA 77, 7347-7351.

18. Figurski, D.H. and Helinski, D.R. (1979). Proc. Natl. Acad. Sci. USA 76, 1648-1652.

19. Hirsch, P.R. and Beringer, J.E. (1984). Plasmid 12, 139-141.

20. Berg, D.E., Davies, J., Allet, B. and Rochaix, J. D. (1975). Proc. Natl. Acad. Sci. USA 72, 3628-3632.

21. Ruvkun, G.B. and Ausubel, F.M. (1981). Nature 289, 85-88.

22. Ruvkun, G.B. and Ausubel, F.M. (1980). Proc. Natl. Acad. Sci. USA 77, 191-195.

23. Jones. R., Woodly, P. and Robson, R. (1984). Molec. Gen. Genet. 197, 318-327.

24. Kennedy, C., Gamal, R., Humphrey, R. , Ramos, J., Brigle,K. and Dean, D. (1986). Molec. Gen. Genet. 205, 318-325.

25. Avtges, P., Kranz, R. and Haselkorn, R. (1985). Molec. Gen. Genet. 201, 363-369.

26. Corbin, D., Ditta, G. and Helinski, D.R. (1982) J. Bacteriol. 149, 221-228.

27. Ruvkun, G.B., Sundaresan, V. and Ausubel, F.M. (1982). Cell 29, 551-559.

A TN5 INDUCED NIFA LIKE MUTANT OF AZOSPIRILLUM BRASILENSE

M. Singh* and W. Klingmüller**
Max-Planck-Institut für Züchtungsforschung, D-5000 Köln 30*
and Lehrstuhl für Genetik, Universität Bayreuth,
D-8580 Bayreuth**, FRG

Summary: Using the suicide vector pGS9, transposon Tn5 mutagensis was employed to isolate three Nif⁻-mutants of A. brasilense strain ATCC29710 (Cd). One of the mutants, Nif27, has been characterized physically as well as biochemically. This mutant is not defective in nifHDK genes. The cloned nif27::Tn5 DNA does not show homology to the nif/gln gene probes of K. pneumoniae or to the fix gene probe of Bradyrhizobium japonicum. The results of western blotting show that Nif27 is unable to synthesize the nitrogenase iron protein and thus this mutant is defective in a nifA like gene. Keywords: Tn5 mutagenesis, Nif⁻-mutant, nifA.

Introduction

Genetical analysis of nitrogen fixation (nif) genes of Azospirillum has been restricted to the nitrogenase structural genes (nifHDK) because these are the only nif genes which show homology to the K. pneumoniae nif genes, thus enabling their cloning and characterization (1, 2, 3, 4). For the isolation of the other nif genes of Azospirillum, transposon mutagenesis offers a valuable method as the mutation is simultaneously tagged, which makes the cloning of the gene a straight forward task. Keeping this in mind, we carried out Tn5 mutagenesis in A. brasilense and isolated 3 different Nif⁻ mutants which could not fix nitrogen at all (5). Here we report on the physical and biochemical characterization of one of these mutants, Nif27.

Azospirillum IV: Genetics, Physiology, Ecology
Edited by W. Klingmüller

Materials and Methods

Bacterial strains and plasmids

The A. brasilense strain ATCC 29710 served as the wild type. For transposon mutagenesis, a rifampicin and streptomycin resistant derivative of 29710 was used. The Nif^- mutant Nif27 is a Tn5 containing derivative of 29710 rif Sm. E. coli strain HB101 was used for cloning work and was obtained from J. Collins, FRG. The following plasmids were used: pSUP202 (6), pSA30 (7), pCK1 (8), pMM12 (9), pGE100 (9) and pMS188 (5).

Media and growth conditions

Complete and minimal media used for Azospirillum and E. coli have been described previously (4). In vivo activity of nitrogenase was measured by C_2H_2 reduction in nitrogen-free, semi-solid MAZ (4). The following concentrations (µg/ml) of antibiotics were used: Chloramphenicol (Cm), 25; Kanamycin (Kan), 40; Tetracyclin (Tc), 15; Rifampicin (rif), 100.

DNA manipulations

Isolation, purification and other manipulations of plasmid and chromosomal DNA were done as described earlier (10, 11).

Results and discussion

The isolation of Tn5 induced mutant Nif27 and the cloning of the mutagenized nifDNA has been described earlier (4, 5).

Growth characteristics of Nif27

Since the Nif^- phenotype may result from a mutation in a non-nif gene, e.g. glnA/ntrBC, which would either result in auxotrophy or renders the mutant incapable of utilizing certain nitrogen sources, the mutant Nif27 was tested in this respect. Nif27 was grown on minimal medium without glutamine. It showed comparable growth to the wild type strain 29710, hence it is not defective in glnA, the structural gene for glutamine synthetase. Furthermore, Nif27 could utilize 1 $mMNH_4^+$, 10 $mMNH_4^+$, histidine, arginine, proline and nitrate when provided as sole sources of nitrogen. From these data, it could be concluded that Nif27 is not defective in ntr genes.

Search for homology with heterologous gene probes

Conclusive homologies (Southern hybridization) have been demonstrated earlier between Azospirillum genomic DNA and the following heterologous gene probes: KpnifHDK (1), KpglnA and fixA of B. japonicum (12, 13). We used these probes to detect homology with the cloned nif27::Tn5 DNA present in plasmid pMS188 (5). No homology was observed with any of the probes showing conclusively that Nif27 is not defective in nifHDK, glnA or fixA type gene.

Western blotting and immunodetection of the iron protein (Ab2) of nitrogenase

Although the nitrogenase structural genes nifHDK are intact in the mutant Nif27, it is possible that the nitrogenase proteins are missing, for example, when the Tn5 insertion had taken place in a regulatory gene similar to nifA in K. pneumoniae which is necessary for the activation of the rest of the nif genes.

Proteins from the crude extracts of the derepressed cultures of the mutant Nif27 and the wild type strain 29710 were separated on SDS-polyacrylamide gels, transferred onto nitrocellulose paper and probed with anti-Rhodospirillum rubrum iron protein-antibodies (anti-Rr2) as described by Hartmann et al. (14). See fig. 1.

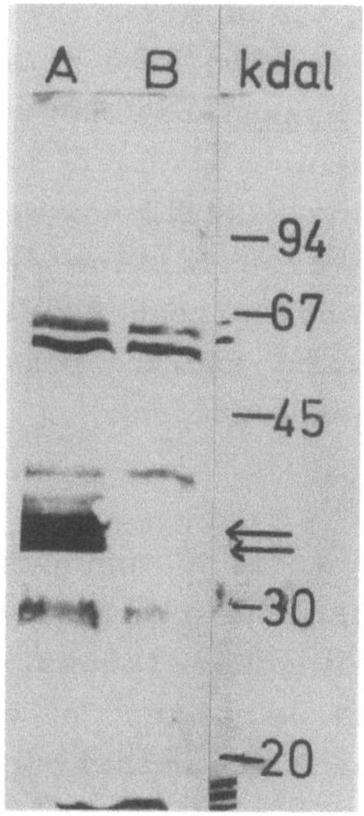

Figure 1. Immunoblotting of the nitrogenase iron proteins of A. brasilense wild type 29710 (A) and mutant Nif27 (B). The arrows show the two bands of the wildtype iron protein (doublet), which are missing in the mutant.

The two strong bands of 36 kdal and 39 kdal precipitated from the wild type strain (fig. 1, lane A) represent the iron protein of A. brasilense (14, 15). These two bands were missing completely in the crude extracts prepared from ammonium grown cultures. It is clear from fig. 1 (lane B) that mutant Nif27 lacks the iron protein of nitrogenase. The iron proteins of A. brasilense and A. lipoferum usually appear as double bands on SDS-polyacrylamide gels (14, 15), which is most probably due to the modification of the iron-protein by the inactivating enzyme (15). The other faint bands precipitated by anti-Rr2 were also observed in the case of ammonium grown cultures and hence are not Nif-specific.

Nair et al. (15) also observed multiple, non-specific bands in A. brasilense probed with anti-Kp2.

The results presented above strongly suggest that the Tn5 induced mutant Nif27 is defective in a nifA type gene. The existence of such a gene in A. brasilense has also been indicated by Pedrosa and Yates (16). Since we have cloned the mutagenized nifDNA from mutant Nif27, it should now be possible to clone the functional nifA gene from Azospirillum spp. Our recent results (Southern hybridization) show strong homology between the EcoRI fragment containing nif27::Tn5 and the plasmids of 3 A. brasilense strains (unpublished observations).

Acknowledgements

We thank Dr. M. Merrick, Prof. H. Hennecke and Prof. A. Pühler for providing the strains. The expert technical assistance of Mrs. M. Ohlraun ist greatly acknowledged. We greatly appreciate the help provided by Dr. A. Hartmann for western blotting. This work was supported by the Bundesministerium für Forschung und Technologie.

References

1. Quiviger, B., Franche, C., Lutfalla, G., Rice, D., Haselkorn, R. and Elmerich, C. (1982). Biochimie 64, 495-502
2. Jara, P., Quiviger, B., Laurent, B. and Elmerich, C. (1983). Can. J. Microbiol. 29, 968-972
3. Fahsold, R., Singh, M. and Klingmüller, W. (1985). In Klingmüller, W. (ed.) Azospirillum III: Genetics, Physiology, Ecology. Springer-Verlag, Heidelberg, p. 30-40
4. Singh, M. and Klingmüller, W. (1985). In Klingmüller, W. (ed.) Azospirillum III: Genetics, Physiology, Ecology. Springer-Verlag, Heidelberg, p. 20-29
5. Singh, M. and Klingmüller, W. (1986). Mol. Gen. Genet. 202, 136-142
6. Simon, R., Priefer, U. and Pühler, A. (1983). Biotechnology 1, 784-791
7. Cannon, F.C., Riedel, G.E. and Ausubel, F.M. (1979). Mol. Gen. Genet. 174, 59-66
8. Kennedy, C. and Robson, R. (1983). Nature 301, 626-628

9. Espin, G., Alvarez-Morales, A., Cannon, F.C., Dixon, R. and Merrick, A. (1982). Mol. Gen. Genet. 186, 518-524
10.Singh, M., Kleeberger, A. and Klingmüller, W. (1983). Mol. Gen. Genet. 190, 373-378
11.Singh, M. and Klingmüller, W. (1985). In Sinha, U. and Klingmüller, W. (eds.) Trends in Molecular Genetics. Spectrum Publishing House, Patna, Delhi, p. 15-34
12.Fogher, C., Bozouklian, H., Bandhari, S.K. and Elmerich, C. (1985). In Klingmüller, W. (ed.) Azospirillum III: Genetics, Physiology, Ecology. Springer-Verlag, Heidelberg, p. 41-51.
13.Fogher, C., Dusha, I., Barbot, P. and Elmerich, C. (1985). FEMS Microbiol. Lett. 30, 245-249
14.Hartmann, A., Fu, H. and Burris, R.H. (1986). J. Bacteriol. 165, 864-870
15.Nair, S.K., Jara, P., Quiviger, B. and Elmerich, C. (1983). In Klingmüller, W. (ed.) Azospirillum II: Genetics, Physiology, Ecology. Experientia Supplementum 48, Birkhäuser-Verlag, Basel, p. 29-38
16.Pedrosa, F.O. and Yates, M.G. (1984). FEMS Microbiol. Lett. 23, 95-101

Tn5-MUTAGENESIS IN *AZOSPIRILLUM BRASILENSE*

M. Vanstockem, A. Milcamps, K. Michiels, J. Vanderleyden and A.P. Van Gool
F.A. Janssens Memorial Laboratory for Genetics
Catholic University of Leuven, Willem de Croylaan 42, B-3030 Heverlee

Summary: A system is described for transposon mutagenesis of *Azospirillum brasilense* using the mobilizable plasmid pSUP2021. Mutants were selected on rich medium containing kanamycin and streptomycin. All the auxotrophs isolated were impaired in their sulfur metabolism. Calcofluor-dark Tn5 mutants were tested for their ability to form flocs.

Keywords: *Azospirillum brasilense*, Tn5 mutagenesis, auxotrophy, Cal^- mutants

Introduction

Since the initial report by Elmerich and Franche (1) on Tn5 mutagenesis of bacteria of the genus *Azospirillum*, various groups were able to isolate Tn5 mutants of *A. brasilense* (2, 3). In these studies, the suicide vectors of Selvaraj and Iyer (4) and Simon *et al.* (5) were used. *Azospirillum* strains mutagenized sofar were all ATCC strains and antibiotic resistant derivatives thereof.

In our present study, we attempted to isolate Tn5 induced mutants of *A. brasilense* strain Sp245. Strain Sp245 was isolated from the root interior of wheat in Brasil and described by Baldani *et al.* (6). This strain reproducibly invades the roots of wheat plants and is therefore of particular interest to study *Azospirillum*-plant interactions, whereas inoculation effects on nitrogen incorporation varied from experiment to experiment. We have shown that the genome of this strain contains sequences homologous to the chromosomal virulence region of *Agrobacterium tumefaciens*, as was shown previously for strain Sp7 (7; see also Eyers *et al.*, this volume).

Azospirillum IV: Genetics, Physiology, Ecology
Edited by W. Klingmüller

Materials and methods

Bacterial strains and plasmids

The bacterial strains and plasmids used in this study are listed in Table I.

Table I. Bacterial strains and plasmids

Strain	Relevant properties	Reference and/or source
E. coli		
SM10	*thi, thr, leu, recA* [RP_4 2-Tc::Mu] integrated, Km^r	Simon *et al.* (5)
HB101	r^-_B, m^-_B, *recA*	Maniatis (10)
Azospirillum brasilense		
Sp7	Ap^r	ATCC29145
Sp7 Rif^r	Ap^r, spontaneous Rif^r derivative of Sp7	This laboratory
7030	Ap^r, NG induced Sm^r derivative of Sp7	C. Elmerich
Sp245	Sp^r	J. Döbereiner
Sp245 Sm^r	Sp^r, spontaneous Sm^r derivative of Sp245	This laboratory
Sp245 NR^-	Sp^r, nitrate reductase negative mutant of Sp245	J. Döbereiner
Plasmid		
pSUP2021	*tra* mob^+ Tn*5* donor Ap^r, Cm^r Tc^r, Km^r (Tn*5*)	Simon *et al.* (5)
pRK290	*tra* mob^+ *IncP* replicon, Tc^r	Ditta *et al.* (8)
pRK2013	tra^+ ColE1 replicon, Km^r	Figurski *et al.* (9)

Media and growth conditions

Escherichia coli strain SM10 was maintained on LB agar and grown in LB broth supplemented with kanamycin (25 μg/ml). *Azospirillum* strains were maintained on yeast-extract-peptone (YEP) agar and grown in YEP broth. Tn*5* induced mutants were selected on YEP-agar with 50 μg/ml of kanamycin and 250 μg/ml of streptomycin. Screening for auxotrophs was performed in replicas on a minimal malate medium (MMAB) and a rich medium.

Bacterial matings

Azospirillum acceptor strains and *E. coli* donor strains were mixed in a certain ratio, after dilution or concentration of the cells (see Table III). The conjugation mix was spread on a nitrocellulose filter, placed on YEP agar and incubated at 30°C for 6 up to 24 hours in a humid atmosphere.

In the case of conjugations with preincubation, the *Azospirillum* acceptor strain was spread on the nitrocellulose filter and incubated for 2 h at 30°C before being overlayed with the *E. coli* donor strain.

DNA isolation

Plasmid DNA of *E. coli* was prepared by using the alkaline lysis method, described in Maniatis (10). *Azospirillum* plasmid DNA was isolated by the rapid detection method of Kado and Liu (11).

DNA hybridization

Colony hybridization was done as described by Maniatis *et al.* (10) using [α-^{32}P]-dCTP labeled probes of pSUP2021 obtained by using the nick translation kit of Amersham.

Southern blots were obtained and hybridized as described by Silhavy *et al.* (12).

Filters were autoradiographed at -80°C by using Fuji RX films and intensifying screens (Kyokko special).

Screening for auxotrophs

Colonies of Km^rSm^r exconjugants were replica plated on MMAB-medium. Auxotrophs were submitted to a characterization test as described by Davis and Botstein (13).

Calcofluorwhite test and floculation

Colonies of Km^rSm^r exconjugants were replica plated on YEP medium containing 0.02 % calcofluorwhite. Screenings were performed by irradiation with U.V. light. Calcofluor-dark mutants were further tested for their ability to form flocs as it was described by Sadasivan *et al.* (14).

Results and discussion

Transposon Tn*5* mutagenesis. Previously, we have demonstrated that in crosses between S17-1 (pSUP2021) and *A. brasilense* Sp7, Tn*5* induced kanamycin resistant mutants of *Azospirillum* can be obtained with a frequency of 10^{-7} per recipient cell (3).

At first, we wanted to compare different strains of *A. brasilense* (see Table I) with respect to Tn*5* mutagenesis. *E. coli* S17-1 (containing pSUP2021) was conjugated with *Azospirillum* strains (see Materials and Methods for details). Km^rSm^r exconjugants were selected on rich medium with 50 μg/ml of kanamycin and 250 μg/ml of streptomycin. We have shown earlier (3) that Tn*5* confers kanamycin and streptomycin resistance to *Azospirillum*. Results are shown in Table II.

Table II. Tn*5* mutagenesis of *A. brasilense* strains

Crosses (a)		Frequency of Km^rSm^r exconjugants
S17-1(pSUP2021)	x Sp7	10^{-7}
	x 7030 (b, c)	9×10^{-7}
	x Sp7 Rif^r (b)	5×10^{-8}
	x Sp245	8×10^{-8}
	x $Sp245Sm^r$ (c)	10^{-7}
	x $Sp245NR^-$	10^{-7}

(a) Crosses were done on membrane filters with *E. coli* cells grown to a density of 3 x 10^7 cells per ml and *Azospirillum* cells grown to a density of 1 x 10^8 cells per ml.
(b) These strains have lost the 115 Mdalton plasmid of the parent strain Sp7.
(c) These strains are resistant to streptomycin (250 µg/ml).

Repeatedly, it was observed that strain 7030 gives the highest number of Km^rSm^r exconjugants among the tested strains. Strain 7030 is a mutant of strain Sp7, obtained by chemical mutagenesis (C. Elmerich), that differs from the parent strain through (i) its resistance to streptomycin and (ii) loss of the 115 Mdalton plasmid. From our results it can be concluded that none of these properties appears to have any influence on the frequency with which Tn*5* mutants can be obtained. We also tested the various *Azospirillum* strains as recipients for the broad host range plasmid pRK290 in crosses between HB101 (pRK290) x HB101 (pRK2073) x *Azospirillum.* Transfer of the plasmid pRK290 from *E. coli* to *Azospirillum* works almost equally well for all *Azospirillum* strains, with strain Sp245 and its derivatives even slightly superior. Next we wanted to try varying mating conditions in an attempt to increase the frequency of Tn*5* transposition in the *Azospirillum* genome. The results of crosses between SM10 (pSUP2021) and *A. brasilense* 7030 are shown in Table III.

Table III. Tn*5* mutagenesis of *A. brasilense* 7030

Mating experiment	OD of mating cells		Ratio	Frequency of Tn*5* transposition	
	SM10	7030	Donor:Recipient	(a)	(b)
1	0.2	0.3	1:1	10^{-5}	10^{-5}
2	0.35	0.7	1:2	/	10^{-5}
3	0.6	1.24	1:10	$4x10^{-7}$	$2x10^{-6}$
4	0.8	1.4	1:20	$5x10^{-8}$	$4x10^{-6}$

(a) Donor and recipient cells are mixed before applying to the conjugation filter

(b) Recipient cells are preincubated for 2 hrs on the conjugation filter and subsequently overlayed with donor cells.

These results demonstrate that frequencies as high as 10^{-5} can be obtained. Secondly, it appears that the growth stage of the recipient cells appears not to influence the transposition frequency as long as the recipient cells are preincubated on the conjugation filters. On the contrary, when donor and recipient cells are applied at the same time on the conjugation filter, Tn*5* transposition decreased drastically with increasing age of the cells. These mating conditions, i.e. preincubation of recipient cells, were subsequently tested for strains Sp7 and Sp245. Under no circumstances a Tn*5* transposition frequency of 10^{-5} can be obtained with these strains. Sofar, we have no experimental data to explain the differences in Tn*5* transposition frequency for the tested strains. Since strain 7030 is the only strain that has been obtained by chemical mutagenesis, we can only suggest that in this strain a gene or a set of genes has been mutated whereby the activities of the corresponding wild type alleles suppress Tn*5* transposition.

Isolation and characterization of Tn*5*-induced auxotrophic and Cal^- mutants. Km^rSm^r exconjugants of strain 7030 were replica plated on minimal medium. Of the 700 presumptive Tn*5*-containing strains tested, 5 were auxotrophs, a frequency of 0.7 %. All these mutants were shown to contain Tn*5* by colony hybridization. Three of the auxotrophs were Cys^-, one was Met^-, and one was deficient in a more general sulfur metabolism function (supplemented by methionine, cysteine and thiosulfate). This unusual distribution of Tn*5*-induced auxotrophs was also observed, although less extreme, by Meade *et al.* (15) for *Rhizobium meliloti* and by Elmerich (personal communication) for *A. brasilense*. It is assumed that it reflects a high number of genes involved in sulfur metabolism.

Km^rSm^r exconjugants were subsequently tested for calcofluor white staining. It was shown previously (K. Michiels, personal communication) that *Azospirillum* can be stained with calcofluor white, a specific strain for β-linked glucans (16), present in e.g. exopolysaccharides. Of 1500 presumptive Tn*5*-containing strains tested, 3 showed a calcofluor white minus (Cal^-) phenotype.

Subsequently it was shown that these mutants were not able to produce flocs (floc$^-$). It is currently under investigation whether the phenotype can be restored by complementing the Cal$^-$ mutants with a cosmid library of *A. brasilense*.

Physical analysis of Km^rSm^r exconjugants of strain Sp245

As was mentioned in the introduction, strain Sp245 is of particular interest in studying the *Azospirillum*-plant interaction because of its ability to penetrate the roots of wheat.

When we used the optimum mating conditions as for strain 7030, no enhancement of the transposition frequency could be detected (as was the case for strain Sp7). This was also the case when we varied mating temperature, cell ratio or growth stage.

Nevertheless we obtained some Km^rSm^r exconjugants (freq. 10^{-7}) for further analysis. DNA of 5 randomly picked Km^R colonies was separated on an agarose gel after Kado and Liu lysis. It was blotted on nitrocellulose and hybridized with a ^{32}P-labeled Tn*5* probe. In all 5 cases Tn*5* seemed to have inserted in the largest megaplasmid of Sp245. More hybridization studies on total digested DNA will allow us to determine the exact spot of Tn*5* insertion and make further speculations about the randomness of the Tn*5* transposition process in Sp245.

References

1. Elmerich, C. and Franche, C. (1982). *In* W. Klingmüller (ed.). *Azospirillum*: Genetics, Physiology, Ecology. Exs 42, Birkhaüser, Basel
2. Singh, M. and Klingmüller D.G. (1986). Mol. Gen. Genet. *202*, 136-142
3. Vanstockem, M., Michiels, K., Vanderleyden, J. and Van Gool, A. (1987). Appl. Env. Microbiol. *53*, 410-415
4. Selvaraj, G. and Iyer, V.N. (1983). J. Bacteriol. *156*, 1292-1300
5. Simon, R., Priefer, U. and Pühler, A. (1983). Bio/Technology *1*, 784-790
6. Baldani, V.L.D., Alvarez, M., Baldani, J.I. and Döbereiner, J. (1986). Plant and Soil *90*, 35-46
7. Michiels, K., Maris, M., Vanstockem, M., Vanderleyden, J. and Van Gool, A. (1985). Arch. Intern. Physiol. Biochim. B153

8. Ditta, G., Stanfield, S., Corbin, D. and Helinski, D. (1980). Proc. Natl. Acad. Sci.. USA *77*, 7347-7351
9. Figurski, D. and Helinski, D. (1979). Proc. Natl. Acad. Sci. USA *76*, 1648-1652
10. Maniatis, T., Fritsch, E. and Sambrook, J. (1982). Molecular cloning, a laboratory manual. Cold Spring Harbor Laboratory, N.Y.
11. Kado, C.I. and Liu, S. (1981). J. Bacteriol. *145*, 1365-1373
12. Silhavy, T., Berman, M. and Enquist, L. (1984). Experiments with gene fusions. Cold Spring Harbor Laboratory, N.Y.
13. Davis, R., Botstein, B. and Roth, J. (1980). A manual for genetic engineering. Advanced bacterial genetics. Cold Spring Harbor, N.Y.
14. Sadasivan, L. and Neyra, C. (1985). J. Bacteriol. *1631*, 716-723
15. Meade, H., Long, S., Ruvkun, G., Brown, S., Ausubel, F. (1982). J. Bacteriol. *149*, 114-122
16. Hughes, J. and McCully, N.E. (1975). Stain Technol. *50*, 319-329

ISOLATION OF TRANSPOSON INDUCED AUXIN-NEGATIVE MUTATIONS IN AZOSPIRILLUM LIPOFERUM

Mohammed Safwat Abdel-Salam[a] and Walter Klingmüller
Lehrstuhl für Genetik, Universität Bayreuth, Universitätsstraße 30, D-8580 Bayreuth, Federal Republic of Germany

Summary: A highly effective method for transposon Tn5 mutagenesis in A. lipoferum is reported. Tn5 was delivered to A. lipoferum using the suicide plasmid pSUP2021 and the donor strain E. coli SM10. Neomycin-resistant transconjugants were detected at a frequency of $6x10^{-6}$/recipient. Different types of mutants were isolated, e.g. auxotrophs, coloured, IAA-negative, and IAA-overproducers. Two hot spots were found in cysteine and methionine biosynthesis genes. Tn5-induced mutants were found to be relatively stable, with reversion rates of $2 - 20 \times 10^{-8}$. A gene which is a part of the carotenoid pathway is closely linked to the histidine genes. Key words: Transposon mutagenesis, Tn5, indoleacetic acid.

Introduction

Azospirilla seem to have the potential to be developed as biofertilizers for their abilities to fix nitrogen and to excrete phytohormones, e.g. indole 3-acetic acid (IAA), combined with a tendency to associate with the roots of many grain crops (1).

The shortage of Azospirillum-mutants, due to the difficulty in mutation induction (2, 3), is one of the main problems in Azospirillum studies. Since the report by Elmerich and Franche (4) on Tn5-induced auxotrophs in A. brasilense 7000, different protocols have been suggested to improve the efficiency of

a) Present address: Gene-Technology Lab., National Research Centre, El-Tahrir Street, Dokki, Cairo, Egypt.

Azospirillum IV: Genetics, Physiology, Ecology
Edited by W. Klingmüller

transposon mutagenesis in Azospirillum. Using the suicide plasmid pSUP2021, Tn5 could be introduced into A. brasilense and A. lipoferum with frequencies of 10^{-8} per recipient (5, 6). With suicide plasmid pGS9 frequencies of 1 - 5 x 10^{-6} were obtained in A. brasilense (6). No comparable efficient method has been yet reported for A. lipoferum.

The purpose of this study was to establish an efficient transposon-system for inducing mutations in A. lipoferum. We were also interested in isolating IAA-mutants to increase the knowledge of the involvement of Azospirillum excreted IAA in plant-Azospirillum interactions.

Materials and methods

Bacterial strains and plasmids: E. coli SM10 and the suicide plasmid pSUP2021 (7), carrying the Km^r-Nm^r Tn5, were used to deliver the transposon to A. lipoferum ATCC 29708, Rif^r, Sm^r, (this lab.).

Media and growth conditions: Complete medium for E. coli and Azospirillum was Luria broth (LB) pH 7.0. Minimal medium for Azospirillum was (MAZ) (8). Medium for bacterial matings was M56 (9) supplemented with 3 g glucose, 5 g malate, 8 mg thiamine, 100 mg threonine and 100 mg leucine per l distilled water.

The following concentrations of antibiotics were used (µg/ml): carbenicillin (Cb) 100, chloramphenicol (Cm) 35, neomycin (Nm) 200; rifampicin (Rif) 100, streptomycin (Sm) 200, and tetracycline (Tc) 15. Auxanographic tests were performed as described by Holliday (10).

Bacterial mating and selection of transconjugants: The E. coli donor strain was incubated overnight at 37° C in LB medium with Nm, Cm and Cb under continous shaking, then diluted tenfold in the same medium and grown another 5 h (log phase). A. lipoferum was incubated overnight at 30° C in LB medium with Rif and Sm while shaking. Donor and recipient cells were concentrated by centrifugation, the pellets washed twice and resuspended in saline. A mixture of donor cells ($5x10^7$) and A. lipoferum cells ($1x10^9$) was plated onto solid M56 medium. The plates were incubated overnight at 30° C and the bacteria collected in 1 ml saline. One hundred ul of the conjugation mixture was plated onto LB with Rif, Sm and Nm to select for A. lipoferum transconjugants. We purified transposon-induced mutants after an incubation of 3-4 days at 30° C by twice streaking them out on the selective medium to obtain single colonies.

Measurement of IAA: The colourimetric Salkowski reaction (11) was used to measure the amount of IAA excreted by A. lipoferum and the transconjugants.

DNA isolation: Total genomic DNA was isolated as in (12), while plasmid DNA was isolated as described in (13).

DNA manipulations: DNA digestion with restriction enzymes, gel electrophoresis, DNA staining, and Southern transfer were as in (14). DNA hybridization, posthybridization and autoradiography were as in (15).

Results

The effect of donor to recipient ratio on transconjugation rate:

We tried to optimize the donor-recipient ratio. Under the conditions described in materials and methods, we found an optimum of 1 donor: 10-20 recipients. More than 600 transconjugants were obtained / plate (ca. $6x10^{-6}$/recipient) when 1 x 10^9 cells of A. lipoferum were conjugated with 5 x 10^7 E. coli SM10 donor strain cells.

Transposon mutants:

More than 5000 Nm^r transconjugants were obtained, and tested for different types of mutations. We could identify about 80 auxotrophs 4 red mutants, 11 IAA-negative mutants, which produce IAA with about 8 to 55 % efficiency of the wild type, and 8 IAA-overproducers, producing 1,5 times and more IAA than the wild-type. We also detected plasmid co-integration with a frequency of $2x10^{-6}$ as judged by retention of the antibiotic resistance markers encoded by the plasmid, Cb^r, Cm^r, and Tc^r. Using the Holliday test (10), the requirements of the auxotrophs were determined as: Adenine, histidine, tryptophan, methionine, cysteine, leucine, thiamine, cytosine, or pantothenic acid. We found an excess of cysteine and methionine auxotrophs (50% and 19% of the auxotrophs, respectively).

Stability of the mutants:

Four auxotrophs were chosen to study their reversion, viz. Al 58 (his requirement), Al 242 (his), Al 171 (trp), and Al 401 (cys). They were plated on MAZ medium and incubated for 5 days, the colonies coming up (revertants) were counted. The rates of reversion were 172, 17, 39, and 72 x 10^{-9} / per cell, respectively. All revertants found were Nm-sensitive; hence, no transposon was left. This was confirmed by randomly chosing three revertants from each mutant for Southern blotting and hybridization to a radiolabeled Tn5-Hind III fragment serving as probe. Interestingly, 10% of 102 revertants of the histidine mutant Al 58 were red coloured indicating a defect in the carotenoid biosynthesis in those revertants due to transposon excision.

Isolation of IAA mutants:

The excretion of IAA by the transconjugants was compared with that of the wild-type. Both IAA-negative and IAA-overproducer mutants could be identified. All IAA mutants exhibited normal growth. Therefore, differences in IAA production should not be due to growth abnormality. The detected amounts of IAA excreted by 14 different IAA mutants, among them 3 overproducers, following 48, 72, and 96 h incubation are presented in Table 1 (see end of this article). The maximum rate of IAA excretion in all strains occurred on the second day of incubation. Thin layer chromatographic analysis was used as described in (11) to identify IAA and related substances in the supernatant of IAA-negative mutants. The results showed that some IAA remained, although much less than in the wild-type.

Physical analysis of Tn5-mutants:

The copy numbers and location of Tn5 in the genome was studied in several mutants. Total DNA was isolated, digested with *Eco* RI, the fragments separated by gel electrophoresis and blotted onto nitrocellulose filters. These were hybridized against ^{32}P-Tn5-*Hind* III fragment. Since Tn5 has no *Eco* RI site, each hybridization signal represents one copy of Tn5. The results revealed that Tn5 was randomly inserted and only a single copy was present per genome of each mutant. The physical analysis of some of the IAA-negative mutants is presented in Fig. 1 (for explanation see following page):

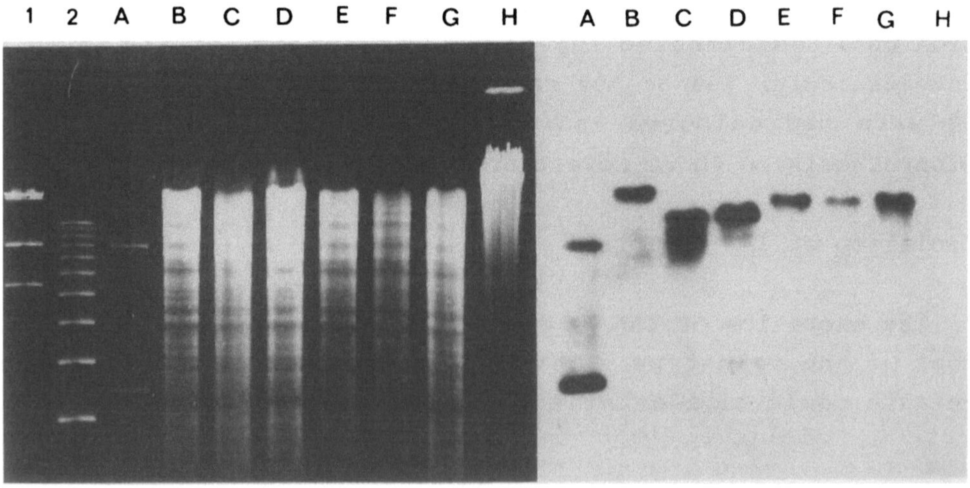

Fig. 1. Physical analysis of Tn5 in some IAA-negative mutants. At left is a 0.7 % agarose gelelectrophoresis of (1) Lambda DNA digested with HindIII; (2) kilobase ladder; pSUP2021 HindIII digested; (B to G) DNAs digested with EcoRI of Al iaa 128, Al iaa 204, Al iaa 140, Al iaa 37, Al iaa 106, Al iaa 320; (H) wild-type *A. lipoferum*. At right is the corresponding autoradiogram after hybridization with ^{32}P-Tn5-HindIII fragment as probe.

Discussion

In the present study we describe a system for inducing Tn5 mutations in *A. lipoferum* using a plate mating. The results establish the importance of the donor to recipient ratio for obtaining high transconjugation rates. They also demonstrate that Tn5 induced stable mutants in *A. lipoferum* can be obtained with only one Tn5 copy randomly inserted per genome. This will facilitate further analysis of the tagged genes.

Among the auxotrophic mutants obtained, there was an excess of methionine and cysteine auxotrophs which, for the latter has previously been reported with Tn5 mutagenesis in *E. coli* K-12 (16). The same effect was found with Tn5 mutagenesis in *Pseudomonas aeruginosa* for cysteine and methionine auxotrophs (17). This finding could indicate the existence of hot spots in the genome for Tn5 mutagenesis of the genes involved.

While studing the revertants of the histidine auxotroph Al 58, we found that 10% of 102 revertants were red coloured. This fact, together with the relatively high reversion frequency of this mutant (ca 2×10^{-7}), could be interpreted as an indication that a suppressor mutation in the carotenoid biosynthesis genes is responsible for the reversion. However, since all reversions had lost the resistance to Nm, revealing the excision and loss of Tn5, this is unlikely. Rather, it can be supposed that in *A. lipoferum*, some genes of both histidine and carotenoid biosynthesis are linked and that Tn5 in mutant Al 58 is located in this region. Then, upon the Tn5-excision, a deletion or localized rearrangement could result in a red coloured revertant.

As to the IAA-negative mutants, their normal growth behaviour is an indication that Tn5 had inserted in the biosynthesis genes of IAA. However, all of our IAA negative mutants had a residual IAA biosynthesis. Even the tightest one still excreted 8.6% IAA after 96 h, compared to the wild-type. Since the insertion of a transposon into a given gene leads to non-leaky polar mutation (18) we conclude that at least two different pathways for IAA biosynthesis, or more than one copy of the genes involved in one pathway are present in *A. lipoferum* 29708.

The high rate of transconjugation reported here with *A. lipoferum* and the efficiency of transposon-mutation induction should greatly facilitate the cloning and mapping of the respective genes, e.g. the genes for IAA biosynthesis and its regulation.

Acknowledgements

We thank Mrs. J. Trnková for technical assistance and Mrs. B. Gubitz for typing the manuscript. This work was supported by the Bundesministerium für Forschung und Technologie.

References

1. Okon,Y.(1985). Trends in Biotechnology 3, 223-228
2. Wood, A.G., Menezes, E.M., Dykstra, C., Duggan, D.E. (1982). In: Klingmüller, W. (ed.) Azospirillum: Genetics, Physiology, Ecology. Experientia Supplementum 42, Birkhäuser-Verlag, Basel, pp. 18-34
3. Del Gallo, M.M., Gratani, L., Morpurgo, G. (1985). In: Klingmüller, W. (ed.) Azospirillum III: Genetics, Physiology, Ecology. Springer-Verlag, Heidelberg, pp. 85-97

4. Elmerich, C., Franche, C. (1982). In: Klingmüller, W. (ed.) Azospirillum: Genetics, Physiology, Ecology. Experientia Supplementum, 42, pp. 9-17
5. Vanstockem, M., Michiels, K., Vanderleyden, J., Van Gool, A. (1985). In: Klingmüller, W. (ed.) Azospirillum III: Genetics, Physiology, Ecology. Springer-Verlag, Heidelberg, pp. 74-84
6. Singh, M., Klingmüller, W. (1986). Mol. Gen. Genet. 202, 136-142
7. Simon, R., Priefer, U., Pühler, A. (1983). Bio/Technology 1, 784-791
8. Albrecht, S.L., Okon, Y. (1980). In San Pietro, A., (ed.): Methods in Enzymol. 69, pp. 740-749. Academic Press, New York
9. Nguyen, N.D., Göttfert, M., Singh, M., Klingmüller, W. (1983). Mol. Gen. Genet. 192, 439-443
10. Holliday, R. (1956). Nature 178, 987
11. Hartmann, A., Singh, M., Klingmüller, W. (1983). Can. J. Microbiol. 29, 916-923
12. Chesney, R.H., Scott, J.R., Vapnek, D. (1979). J. Mol. Biol. 130, 161-173
13. Humphreys, G.O., Willshaw, G.A., Anderson, E.S. (1975). Biochim. Biophys. Acta. 383, 457-463
14. Maniatis, T., Fritsch, E.F., Sambrook, J. (1982). Molecular cloning, a laboratory manual. Cold Spring Harbor.
15. Schwabe, G., Klingmüller, W. (1985). In: Klingmüller, W. (ed.) Azospirillum III: Genetics, Physiolgy, Ecology. Springer-Verlag, Heidelberg, pp. 52-62
16. Shaw, K.J., Berg, C.M. (1979). Genetics 92, 741-747
17. O'Hoy, K., Krishnapillai, V. (1985). FEMS Microbiol. Lett. 29, 299-303
18. Kleckner, N., Roth, J., Botstein, D. (1977). J. Mol. Biol. 166, 125-159

Table 1. Excretion of IAA[a)] by A. lipoferum and its IAA-mutants

Strain	IAA (µg/ml)[b)]		
	48h	72h	96h
a) Wild-type			
A. lipoferum 29708	19.8	28.2	30.4
b) IAA-negative			
Al iaa 37	2.5	2.7	2.6
Al iaa 106	3.6	3.3	4.1
Al iaa 128	3.6	3.0	3.3
Al iaa 140	2.5	6.5	6.9
Al iaa 204	3.9	5.5	5.8
Al iaa 320	3.8	4.1	3.8
Al iaa 13	14.5	14.6	16.9
Al iaa 22	6.4	8.8	13.3
Al iaa 36	9.9	8.8	13.3
Al iaa 99	9.9	13.6	16.8
Al iaa 666	9.4	15.4	15.2
c) IAA-overproducers			
Al op 33	52.0	61.0	64.8
Al op 136	56.3	58.0	63.5
Al op 779	37.8	48.3	47.6

a) IAA was measured in the supernatant of MAZ medium with 0.1 mg L-tryptophan/ml, using the Salkowski test. All cultures had the same cell concentration as measured by OD 578.

b) The amounts of IAA excreted represent an average of three repeats.

AROMATIC AMINO ACID AMINOTRANSFERASES OF AZOSPIRILLUM LIPOFERUM AND THEIR POSSIBLE INVOLVEMENT IN IAA BIOSYNTHESIS

E. Ruckdäschel[1], B. Lewis Kittell[2], D.R. Helinski[2] and W. Klingmüller[1]

[1] Lehrstuhl für Genetik, Universität Bayreuth, Universitätsstraße 30, D-8580 Bayreuth, FRG
[2] Department of Biology B-022, University of California San Diego, La Jolla, CA 92093, USA

Summary: Four aromatic amino acid aminotransferases were identified in *A. lipoferum ATCC 29708*. Each enzyme can in vitro catalyze the transamination of tryptophan, which is the initial reaction in the biosynthesis of IAA via the transamination pathway. Evidence is presented that one of these enzymes is induced in cells that are grown to stationary phase in the presence of tryptophan. This may indicate that this enzyme is the real aminotransferase of the transamination pathway in vivo. A strategy to clone the genes of these enzymes is described. Keywords: *Azospirillum*, aromatic amino acid aminotransferases, IAA, complementation.

Introduction

Azospirillum lipoferum produces indoleacetic acid from tryptophan via the transamination pathway (1). In the initial reaction, the amino group of tryptophan is removed by an aminotransferase. Aminotransferases that react with tryptophan can therefore contribute to IAA production. Their specific mutagenesis should lead to IAA^- - mutants. Such mutants will help further the understanding of the role of IAA in the *Azospirillum* - plant interaction.

Azospirillum IV: Genetics, Physiology, Ecology
Edited by W. Klingmüller

Results

Aromatic amino acid aminotransferases in A. lipoferum

In order to identify aminotransferases that react with tryptophan a crude cell extract was run on a polyacrylamide gel under native conditions and stained for aminotransferase activity as described (2).The staining mix contained tryptophan as amino group donor, α-ketoglutarate as acceptor and an oxidoreductant dye (nitroblue tetrazolium) that is reduced during transamination. Dark bands on the gel therefore indicate aminotransferase activity. In a control either the donor or the acceptor for the amino group was omitted. The results of these experiments can be summarized as follows:

1) there are four aromatic amino acid aminotransferases in A. lipoferum that can catalyze the transamination of tryptophan in vitro (fig. 1)
2) all four genes are expressed constitutively; in contrast to E.coli none of them is repressed by tyrosine (fig. 1)
3) one of these enzymes (band D in fig. 1) is most active after prolonged incubation in the presence of tryptophan, indicating that it is the real enzyme of the transamination pathway (fig. 2, part III and IV)
4) different N-sources have no influence on the aminotransferase pattern (fig. 2, part II)
5) different pH values in the staining mix (pH 7.0 - 9.0 was tested) have no influence on the aminotransferase pattern
6) all four enzymes can react with tryptophan, tyrosine and phenylalanine, two of them react also with aspartate, none of the four reacts with isoleucine, leucine or valin (data not shown)
7) the specific activity of all four enzymes in a cell extract is about 30 nmol product formed per min per mg of protein as determined with the spectrophotometric assay of Lin et al. (3)

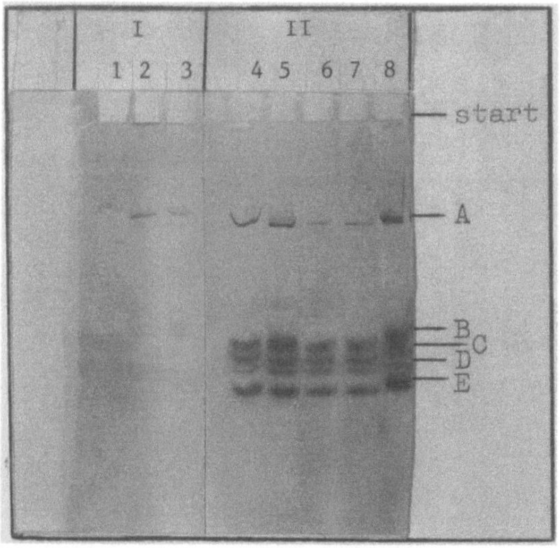

Fig. 1. Analysis of aromatic amino acid aminotransferases in A. lipoferum. The left part of the gel (I) was stained without amino acid (control), whereas the right part (II) was stained for tryptophan aminotransferase activity. Cultures of gel tracks 1 and 5 were grown in LB medium, those of tracks 2,4 and 8 in MAZ medium, those of tracks 3 and 7 in MAZ with 100 µg/ml tryptophan and that of track 6 in MAZ with 100 µg/ml tyrosine. 50 µg of protein were loaded on lane 1 to 7, 100 µg on lane 8. Captital letters B-E indicate the aminotransferase activity bands. The activity of band A is not an aminotransferase.

The in vivo role of these enzymes, however, is not clear. Whereas the third enzyme seems to be the most important for the conversion of tryptophan to IAA, no statement about the metabolic functions of the others can be made. A. lipoferum then, besides for example Rhizobium meliloti, Klebsiella aerogenes or Pseudomonas aeruginosa (4) is another bacterium possessing several aromatic amino acid aminotransferases.

Strategy for the genetic construction of IAA$^-$ - mutants

Since A. lipoferum is reported to possess the transamination pathway only (1) (but note the findings by Abd el Salam and Klingmüller, this volume), IAA$^-$- mutants can be obtained by the blocking of the aminotransferase reaction. This can be done by specific mutagenesis of the aromatic amino acid aminotransferase genes.

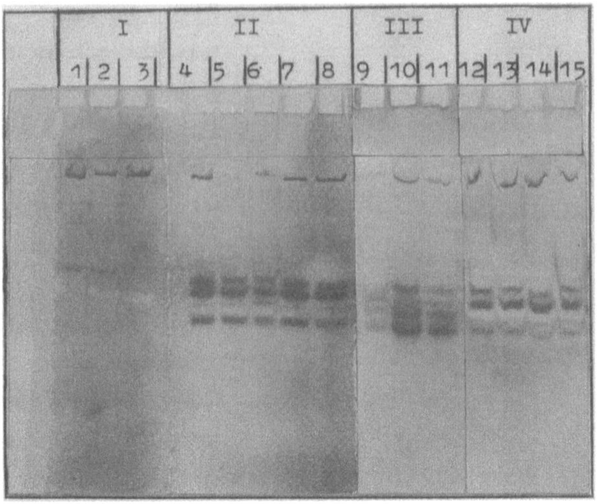

Fig.2. Influence of different N-sources and incubation times on the aromatic amino acid aminotransferase pattern. The gel was stained with tryptophan, except lanes 1 - 3 were stained without amino acid as a control. Part II shows the influence of N-sources: cultures for gel tracks 4, 5, 6, 7 and 8 were grown with NH_4^+, aspartate, tryptophan, phenylalanine and tyrosine as sole N-source, respectively. Part III shows the influence of incubation time (2, 3 and 4 days) in MAZ medium supplemented with 100 µg/ml tryptophan. Part IV shows the pattern of cultures grown for 1, 2, 3 and 4 days in MAZ medium without tryptophan.

The future project therefore can be divided into four steps:

1) cloning the genes of aromatic amino acid aminotransferases
2) mutagenesis of these genes in vitro
3) site directed mutagenesis to exchange the wildtype genes for mutagenized genes
4) determination of the influence of the mutations on IAA production

The genes for aromatic amino acid aminotransferases can be identified in a A. lipoferum gene library by conjugation of the library into E. coli DG 30. This E. coli strain has defects in all E. coli aromatic amino acid aminotransferases (2). It therefore cannot grow on minimal medium without phenylalanine or tyrosine. Since all four enzymes of A. lipoferum can react with tyrosine and phenylalanine, the expression of their genes in E. coli can complement DG30's mutations and allow it to grow without these amino acids.

Two gene libraries of A. lipoferum in mobilizable broad host range cosmids were constructed in the strain E. coli S 17-1. Experiments are under way to identify the aminotransferase genes in these libraries.

References

1. Hartmann, A., Singh, M. and Klingmüller, W. (1983). Can. J. Microbiol. 29, 916-923

2. Gelfand, D.H. and Steinberg, R. (1977). J. Bact. 130, 429-440

3. Lin, E.C.C., Pitt, B.M., Civen, M. and Knox, W.E. (1958). J. Biol. Chem. 233, 668-673

4. Whitaker, R.J., Gaines, Ch.G. and Jensen, R.A. (1982). J. Biol. Chem. 257, 13550-13556

MOBILIZATION OF CRYPTIC PLASMIDS IN AZOSPIRILLUM

M. Pampaluna, M.N. Antonelli, L. Piana, C. Fogher

Istituto di Genetica agraria, Università Cattolica, via E. Parmense 84, 29100 Piacenza, Italy

Summary: In order to facilitate the identification and manipulation of plasmids in *Azospirillum* we attempted to mark them by insertion of the kanamycin-resistant transposon Tn5. Plasmids pACYC184::Tn5-Mob and pBR322 ::Tn5 were used to introduce Tn5 into *A. brasilense* Sp7. A high efficiency of transfer was obtained with pACYC184 containing the Mob region of RP4. All the Sp7 km^r transconjugants isolated were the result of Tn5 transposition to the *Azospirillum* genome. Four per cent of these contained the Tn5-Mob transposon on a replicon mobilizable and stable in *E. coli*. Using a Sp7::Tn5-Mob strain containing the Tn5-Mob on a mobilizable replicon, we tested the mobilization of *A. brasilense* plasmids to other genera. We were able to find km^r transconjugants only with *A. tumefaciens*.

Keywords: *Azospirillum*, cryptic plasmids, mobilization.

Introduction

Very little is known about the physiology and the molecular biology of the association of the *Azospirillum* genus with plants (1). *Azospirillum* harbours plasmids of large molecular weight with which no phenotype has been associated (2). Large plasmids seem to be a common feature in bacteria that associate with plants (3,4,5), and in the particular case of *Rhizobium* they carry information for symbiotic nitrogen fixation (6,7). Plasmid location for *nif* genes was not observed in *Azospirillum* (8) but functions related to bacteria-plant association might be present on its plasmids. *Azospirillum* plasmids are cryptic and therefore it would be very useful to have a marker to study plasmid mobilization and properties. In this paper

Azospirillum IV: Genetics, Physiology, Ecology
Edited by W. Klingmüller

we describe a tagging technique that makes it possible to study the mobilization of the Azospirillum plasmids in other genera.

Materials and methods

Bacterial strains and plasmids

Bacterial strains and plasmids are listed in table 1.

Table 1: Strains and plasmids

Strain	Genotype or phenotype	Reference
A. brasilense		
Sp7	wild type	(2)
E. coli		
S17.1	pro, thi, recA (RP4-2, Tc::Mu-km::Tn7, Tra^{+}, IncP-1)	(9)
C600	trp, his, pro, recA, rif	(11)
HB101	hsd, leu, pro, thi, recA	(10)
DB1706-3	F^{-}, ΔproB-lac, ΔtrpE	D. Berg
A. tumefaciens		
C58C1	C58 cured of pTi58	(12)
R. japonicum		
3I1b143	Rif^{r}, Nal^{r}	U.S.D.A., Beltsville
R. leguminosarum		
LPR1105	Rif^{r} derivative of RCR1001	(13)
Plasmid		
pBR322::Tn5	Tc^{r}, Amp^{r}, Km^{r}	D. Berg
pACYC184::Tn5-Mob	Mob^{+}, Km^{r}, Tc^{r}, Cm^{r}	B. Perroud, (14)
pRK2073	Tra^{+}, Tp^{r}	(15)

Media

LB and minimal used for E. coli have been described (16). Media for Azospirillum were complete medium D (17), and minimal medium K (2). Medium for *Agrobacterium tumefaciens*, *Rhizobium leguminosarum* and *R. japonicum* was YEM (18). Concentration of antibiotics used in media for *E. coli* and *Azospirillum* was 10 µg of tetracycline (Tc), 20 µg ampicillin (Amp), 40 µg kanamycin (Km) and 20 µg of trimethoprim (Tp) per ml; for *A. tumefaciens* and *Rhizobium* was 40 µg of kanamycin, 500 µg of streptomycin and 80 µg of rifampicine per ml.

DNA purification and electrophoretic analysis

DNA of plasmids was isolated by the methods of Eckhardt (19) or Casse *et al.* (20). To detect plasmids, vertical or horizontal agarose gels of varying concentrations were used. Staining of gels and photographic conditions were as described in Maniatis (21).

DNA filter hybridization

After agarose gel electrophoresis, DNA was transferred to nitrocellulose filters using Southern's method (22). Radioactive hybridization probes were labelled in vitro by the nick translation procedure (21).

Genetic techniques

Bacterial matings were performed using standard techniques. Donor and recipient were grown in rich medium at a concentration of about 5×10^8 bacteria per ml. They were mixed in a proportion of 1:9 for *E.coli*: *A.brasilense* and 1:1 for all other conjugation experiments. Triparental matings were in a 1:1:1 ratio. Conjugation mixtures were spread and incubated for 6 h or overnight on rich medium. The bacteria on the plate were then suspended in 2 ml of minimal medium and the appropriate dilutions were spread on selective plates and incubated at 30°C.

Results

Introduction of Tn5 into Azospirillum

Plasmids pACYC184::Tn5-Mob and pBR322::Tn5 were used to introduce Tn5 into *A. brasilense* Sp7. The *E. coli* vector plasmids pACYC184 and pBR322 are neither selftransmissible nor efficiently mobilizable and mantained in *Azospirillum*. The transfer frequency can easily be improved by introducing the RP4-Mob region and doing the conjugation in presence of the transfer functions of plasmid RP4 (23) (table 2). The plasmid pACYC184::Tn5-Mob was constructed by transposition of Simon's (14) Tn5-Mob construct in the vector pACYC184, near the chloramphenicol resistance gene (B. Perroud, unpublished). In strain S17.1 the transacting transfer (Tra^+) function of RP4 is present on the chromosome (23). The strain DB1706-3 does not contain Tra^+ and needs a helper plasmid. The frequency of transfer for both plasmids is shown in table 2. The frequency of kanamycin-resistant derivatives was about 10^{-3} for Tn5-Mob and 10^{-6} for Tn5. A high efficiency of transfer is obtained with pACYC184 containing the Mob region of RP4.

Transposition of Tn5 in Azospirillum

The transposition of Tn5-Mob or Tn5 inside *A. brasilense* Sp7, from the introducing vector to the *Azospirillum* replicons, was assayed through the spontaneous loss of the antibiotic marker of the vector in presence of kanamycin. All the Sp7 transconjugants that went through double single colonies reisolation were sensible to tetracycline, for pBR322::Tn5, or to tetracycline and chloramphenicol, for pACYC184::Tn5-Mob. This is an indication of vector loss and transposition of Tn5 into the *Azospirillum* genome.

Insertion of transposon Tn5 into plasmids

To assay for frequency of Tn5 transposition to *Azospirillum* plasmids a few hundred colonies were purified and used as potential donors in crosses with *E. coli* C600 and S17.1, with selection for transfer of km^r and counter-

Table 2 : Conjugal transfer efficiency of the kanamycin marker between E. coli and A. brasilense Sp7

Exp. n°	Donor	Helper	Recipient	Selected marker	Counterselected marker	Km resistant per recipient
1	S17.1 pACYC184 ::Tn5-mob	-	Sp7	km	pro, thi	2.6×10^{-3}
2	DB1706-3 pBR322 ::Tn5	-	Sp7	km	pro, trp	2.9×10^{-7}
3	DB1706-3 pBR322 ::Tn5	HB101pRK2073	Sp7	km	pro, trp	1.4×10^{-6}

Table 3 : Conjugal transfer of the kanamycin marker between Sp7 and E. coli

Exp. n°	Donor	Helper	Recipient	Selected marker	Counterselected marker	Km resistant (%)
1	Sp7 ::Tn5-mob	-	C600	km	rif	4
2	Sp7 ::Tn5-mob	HB101pRK2073	C600	km	rif	100
3	Sp7 ::Tn5-mob	-	S17.1	km	sm	4
4	Sp7 ::Tn5-mob	HB101pRK2073	S17.1	km	sm	4

Table 4 : Conjugal transfer efficiency of the kanamycin resistance marker between Sp7::Tn5-mob and Agrobacterium and Rhizobium

Exp. n°	Donor	Helper	Recipient	Selected marker	Counterselected marker	Km resistant per recipient
1	Sp7 ::Tn5-mob	-	C58 C1	km	sm	4×10^{-8}
2	Sp7 ::Tn5-mob	HB101pRK2073	C58 C1	km	sm	2.5×10^{-8}
3	Sp7 ::Tn5-mob	-	3I1b143	km	rif	0
4	Sp7 ::Tn5-mob	HB101pRK2073	3I1b143	km	rif	0
5	Sp7 ::Tn5-mob	-	LPR1105	km	rif	0
6	Sp7 ::Tn5-mob	HB101pRK2073	LPR1105	km	rif	0

selection for rifampicine or streptomycin (table 3). The crosses were done using a quick mating procedure (25). Positive transfer was when a confluent growth or five-ten colonies were formed on the point of spotting after three days at 30° C. It is known that transposition of Tn5, in a large number of bacteria, is not site-specific (26). In the experiments with Azospirillum only 4% of the Sp7 km resistant transconjugants (table 3) contain the Tn5-Mob transposon on a replicon mobilizable and stable in E. coli; this comigrate with the 90 Md plasmid of strain Sp7 (2) (figure 1 and 2). The 100% transfer obtained in experiment two (table 3) is inexplicable, unless we suppose a high mobilization efficiency of the chromosome, in the

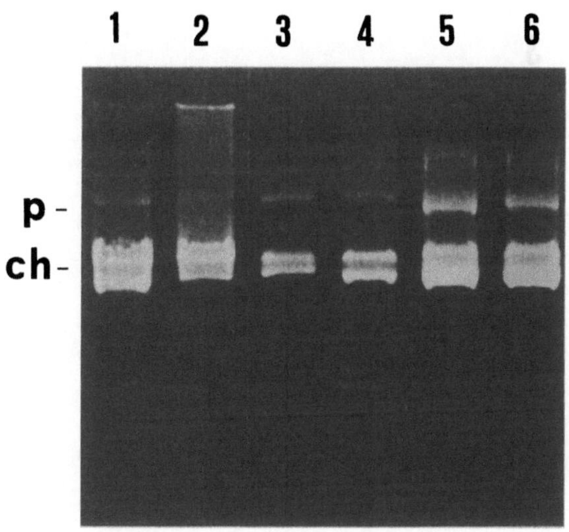

Figure 1. Agarose gel electrophoresis of plasmids from different strains. Lines 1,2,5 and 6, four km^r C600 transconjugants from experiment 1 table 3; lines 3 and 4, two km^r S17.1 transconjugants from experiment 3 table 3. Ch: chromosomal DNA; P: plasmids.

presence of the helper plasmid pRK2073, mediated by Tn5-Mob. Indeed we were unable to prove the presence of plasmids in such C600 km resistant transconjugants. However this is not the case with strain S17.1 (table 3). To check the presence of Tn5 in plasmids of Sp7 transconjugants we did hybridization experiments with ^{32}P-labelled Tn5 probes (figure 2). All Sp7 km^r transconjugants that were able to transfer the kanamycin resistance to E. coli, showed hybridization to labelled Tn5 in correspondence with the 90 Md plasmid (figure 2), the same plasmid probably transferred to E.coli C600 and S17.1.

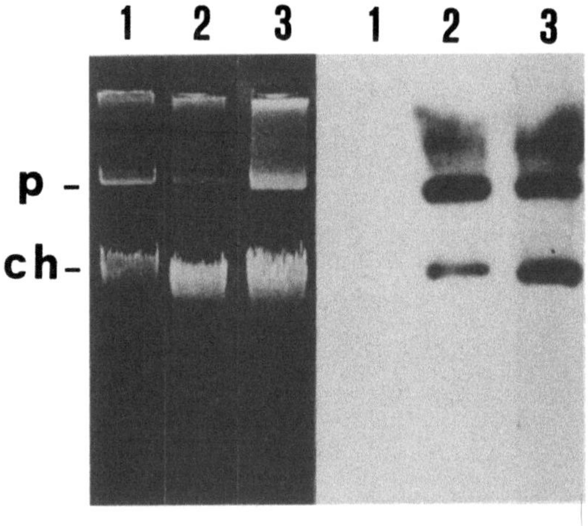

Figure 2. Plasmids gel profile of various A. brasilense Sp7 derivatives and the corresponding autoradiograph of the plasmids gel hybridized with a Tn5 probe. The Tn5 probe was the internal HindIII fragment of Tn5 purified by electroeluition from pBR322::Tn5 digested DNA. Line 1, strain Sp7; line 2, strain C600 km^r from exp. 1 table 3; line 3, strain Sp7::Tn5-Mob from exp. 1 table 2. Ch: chromosomal DNA; P: plasmids.

Mobilization of Azospirillum plasmids to other genera

Using a Sp7::Tn5-Mob strain able to transfer a tagged plasmid to E.coli S17.1, we tested the mobilization of the A. brasilense plasmid inside A. tumefaciens C58C1, R. leguminosarum LPR1105 and R. japonicum 3I1b143. We were able to find km^r transconjugants with C58C1 (table 4). Because of the virtually identical number of C58C1 km^r obtained with and without a helper plasmid, we can suggest that Sp7 contains a cis- or trans-acting transfer function recognised either by the Mob region cloned inside Tn5 or by an analogous function present on the indigenous plasmid.

Discussion

Azospirillum contains plasmids of large molecular weight (2). This seems to be a common feature in bacteria that associate with plants; therefore functions related to bacteria-plant association might be present on Azospirillum plasmids. The recent discovery of the existence of homology between nod regions of Rhizobium and DNA of Azospirillum (24) together with the evidence that this homology is carried, at least partially, on plasmids (C. Elmerich, personal communication; unpublished data from our laboratory) strongly suggests the necessity to analyse Azospirillum for plasmid functions. Azospirillum plasmids are cryptic and, in order to facilitate further analysis in the absence of a readily selectable phenotype, transposon Tn5 was used to label the plasmids. A high rate of transfer was obtained with pACYC184 supply with the Mob region of RP4 cloned on Tn5. The mobilization experiments do not give information on the selftransmissibility of Sp7 plasmids, and this aspect requires more investigation using transposon without the Mob region. To probe exactly which plasmid, with or without deletions, has been transferred in E.coli, it is necessary to test the transferred plasmids with Sp7 plasmid clones through DNA hybridization.

Acknowledgements

M.N.A. was on leave from Enichem Agricoltura; L.P. was the recipient of a postdoctoral fellowship from research contract funds. This work was partially supported by funds from Enichem Agricoltura, Milan.

References

1. Elmerich, C. (1984). Biotechnology 2,967-978
2. Franche, C. and Elmerich, C.(1981). Ann. Microbiol.(Inst. Pasteur) 132A, 3-17
3. Rosenberg, C., Boistard, P., Dénarié, J. and Casse-Delbart, F.(1981). Mol. Gen. Genet. 184,326-333
4. Currier, W.B., Gordon, M.P., Chilton, M.D. and Nester, E.W.(1975). J. Bacteriol. 123,255-264
5. Comai, L. and Kosuge, T.(1980). J. Bacteriol. 143,950-957
6. Banfalvi, Z., Sakanyan, V., Kancz, C., Kiss, A., Dusha, I. and Kondorosi, A. (1981). Mol. Gen. Genet. 184,318-325
7. Rosenberg, C., Boistard, P., Dénarié, J. and Casse-Delbart, F.(1981). Mol. Gen. Genet. 184,326-333
8. Plazinsky, J., Dart, P.J. and Rolfe, B.G.(1983). J. Bacteriol.155,1429-33
9. Simon, R., Priefer, U. and Pühler, A. (1984). Biotechnology 1,784-791
10. Boyer, H.W. and Roulland-Dussoix, D. (1969). J. Mol. Biol. 41,459
11. Hanahan, D. (1983). J. Mol. Biol. 166,577
12. Van Larebeke, N., Genetello, C., Schell, J., Schilperrort, R.A., Hermans, A.K., Hernalsteens, J.P. and Van Montagu, M. (1975). Nature 255,742-743
13. Hooykaas, P.J.J., Snijdewint, F.G.M. and Schilperoort, R.A. (1982). Plasmid 8,73-82
14. Simon, R. (1984). Mol. Gen. Genet. 196,413-420
15. Ditta, G., Stanfield, S., Corbin, D. and Helinski, D. (1980). Proc. Natl. Acad. Sci. USA 77,7347-7351
16. Bozouklian, H., Fogher, C. and Elmerich, C. (1986). Ann. Microbiol. (Inst. Pasteur) 137B,3-18

17. Gauthier, D. and Elmerich, C. (1977). FEMS Microbiol. Letters, 2,101-104

18. Vincent, J.M. (1970). Int. Biology Progr. Hand-book n° 15. Blackwell Pub. Co., Oxford.

19. Eckhardt, T. (1978). Plasmid 1,584-588

20. Casse, F., Boucher, C., Julliot, J.S., Michel,M. and Dénarié, J. (1979). J. Gen. Microbiol. 113,229-242

21. Maniatis, T., Fritsch, E.F. and Sambrook, J. (1982). In "Molecular Cloning". Cold Spring Harbor, New York.

22. Southern, E.M. (1975). J. Mol. Biol. 98,503-517

23. Simon, R., Priefer, U. and Pühler, A. (1983). Biotechnology 2,184-190

24. Fogher, C., Dusha, I., Barbot, P. and Elmerich, C. (1985). FEMS Microb. Letters 30,245-249

25. Franche, C., Canélo, E., Gauthier, D. and Elmerich, C. (1981). FEMS Microbiol. Letters 10,199-202

26. de Bruijn, F.J. and Lupski, J.R. (1984). Gene 27,131-149

A TRANSFORMABLE MUTANT OF AZOSPRILLUM BRASILENSE ATCC 29710

H. El-Khawas and W. Klingmüller
Lehrstuhl für Genetik, Universität Bayreuth,
Universitätsstraße 30, D-8580 Bayreuth, FRG

Summary: Azospirilla are normally not transformable. During our investigations a mutant of A. brasilense ATCC 29710 was isolated, which was transformable with plasmid-DNA (e.g. pSUP106 and pRK290). The mutant also was a better recipient during plasmid-conjugation as compared to the wild type. In consequence the transpositionrate of Tn5 delivered on pSUP 2021 to cells of this mutant was increased as well. Evidence that this mutant is in fact a descendent of strain A. brasilense 29710 was achieved by DNA-hybridization and by analysis of its indigenous plasmid pattern.
Keywords: Azospirillum brasilense, plasmid DNA transformation, conjugation.

Introduction

Nitrogen fixing Azospirilla are associated with roots of several plants in tropical and temperate regions (1,2,3). Genetic studies in Azospirillum will help to identify some of the determinants responsible for the Azospirillum - plant interaction. Methods like DNA-transformation can contribute towards this goal. However, except in the study of Fani et al (4), Azospirilla were found to be not transformable (5). In our investigations a mutant of A. brasilense ATCC29710 was isolated, designated HE1, which, in contrast to the original strain, is transformable.

Azospirillum IV: Genetics, Physiology, Ecology
Edited by W. Klingmüller

Materials and Methods

Bacterial strains and plasmids

They are listed in Table 1.

Media and buffers

All strains were grown in Luria broth (LB) pH 7.0. SOB, SOC and FSB competence buffer were as described by Hanahan (6). The following antibiotics were used (ug/ml): Rifampicin (Rif) 100, streptomycin (Sm) 200, tetracycline (Tc) 10, kanamycin (Km) 100 and chloramphenicol (Cm) 35.

DNA isolation

Plasmid DNA was isolated as described by Rodriguez and Tait (7). Large quantities of plasmid DNA were obtained by the cleared lysate method followed by a final purification in CsCl-EthBr equilibrium density-gradient centrifugation (8). Total genomic DNA was isolated following the method of Chesney et al. (9). Plasmid patterns of *A. brasilense* 29710 and the mutant HE1 were obtained following the Eckhardt method (10).

Transformation

Transformation conditions were according to Hanahan (6) with some modifications (4).

Conjugation

Conjugations were carried out on filter membranes as described before (11).

DNA digestion, electrophoresis, Southern transfer, nick translation, hybridization and autoradiography were as in (12). DNA-DNA hybridization was mainly as described by Johnson et al. (13).

Table 1. Bacterial strains and plasmids

Strains	relevant characteristics	source/ref.
A. brasilense (ATCC)		
29145, 29711,	wild type	
29710,	Spontaneous Rif^r and Sm^r	this lab.
HE1	Transformable mutant of 29710 Rif^r Sm^r	this lab.
A. lipoferum (ATCC)		
29708, 29731	wild type	
Enterobacter cloacae		
MF10	wild type	(19)
E. coli		
HB101	Sm^r, r^-, m^-, $recA^-$	(16)
SM10	thi, thr, leu, sum, Km^r, RP4-2-Tc::Mu	(14)
S17-1	pro, r^-, m^+, RP4-2-Tc::Mu, Km::Tn7, Tp^r, Sm^r	(14)
DH1	F^-, recA1, endA1, gyrA96, thi, hsdR17, supE44, relA1, λ^-	(6)
K12 JC5466	trp, his, recA56, Spc^r	(17)
921	thr, leu, thi, hsr_k, hsm_k, Sm^r	(18)

Plasmids		
pSUP2021	Tra^-, Mob^+, Cm^r, Cb^r, Km^r, (Tn5), pBR325 replicon	(14)
RP4	Amp^r, Tc^r, Km^r	(18)
pRD1	Amp^r, Tc^r, Km^r, His, Nif	(17)
pSUP106	Tc^r, Cm^r	(14)
pRK290	Tra^-, Mob^+, Tc^r, incP1	(20)
pKT210	Mob^+, Cm^r	(21)
pKT231	Mob^+, Km^r	(22)

Results and discussion

Our investigations with A. brasilense were initially concerned with transposon mutagenesis. For our purpose we used the suicide plasmid pSUP2021 containing transposon Tn5 (14). The transposition rate using this plasmid in Azospirillum was only 10^{-7} to 10^{-8} per recipient (15).

In our efforts to obtain higher transposition rates, different mating conditions were tested, e.g. different conjugation media and various temperatures. During these experiments two sizes of colonies were noticed on complete medium (Figure 1). Colonies of average size appeared during 3 days of growth, but some smaller colonies could be found with low rate. We picked 5 small and 5 large bacterial colonies, grew the cells up and mutagenized them with Tn5. Surprisingly one of the small colony sublines gave a mutation rate about 1000 times higher than the other sublines (Table 2).

We then studied conjugative plasmid transfer to this mutant, which was designated HE1, with the conjugative plasmids RP4 and pRD1. The results shown in Table 3 demonstrate that HE1 could be infected by these plasmids, 1000 times better than the wild-type A. brasilense 29710 when E. coli was used as donor strain. From these results we concluded that the high transposition frequency obtained when transferring pSUP2021 onto HE1 was due to an increased conjugation rate.

This encouraged us to test the transformation ability of HE1 with plasmid DNA. Four different wide host range plasmids were tested (Table 4). Transformants were obtained in each case and identified by the antibiotic resistance coded by each plasmid. The parental strain, in contrast, was not transformable.

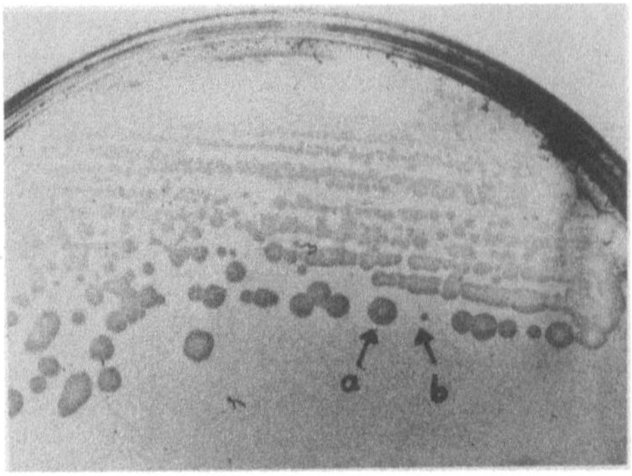

Fig.1: The two colony sizes of A. brasilense 29710 on LB medium. a) common size, b) smaller colony.

Table 2. Transposition frequencies of Tn5 in different *A.brasilense* 29710 sublines

donor	recipient	transposition frequency
E.coli SM10	a) sublines from	
(pSUP2021)	large colonies 1)	1.3×10^{-7}
	2)	8.4×10^{-8}
	3)	2.6×10^{-7}
	4)	1.7×10^{-7}
	5)	9.2×10^{-8}
	b) sublines from	
	small colonies 1)	1.1×10^{-7}
	2)	1.8×10^{-7}
	3)	9.4×10^{-7}
	4)	1.6×10^{-7}
	5)	2.4×10^{-4}

Table 3. Plasmid-conjugation from E. coli to A. brasilense

donor	recipient	conjugation rate*)
E.coli 921	A.brasilense 29710	5.4×10^{-7}
(RP4)	A.brasilense HE1	3.2×10^{-4}
E.coliJC5466	A.brasilense 29710	1.7×10^{-7}
(pRD1)	A.brasilense HE1	1.2×10^{-4}

*) per recipient

Table 4. Transformation of A. brasilense with different plasmids

plasmid	A.brasilense	transformants*)
pRK290	29710	0
	HE1	760
pSUP106	29710	0
	HE1	340
pKT210	29710	0
	HE1	210
pKT231	29710	0
	HE1	240

*) number of transformants per ug plasmid-DNA

To confirm the transformability of HE1 we chose randomly two transformants from both the pRK290 and pSUP106 experiments. Plasmid DNA was isolated by a miniscreen method (6) and the pattern displayed in agarose gels (Figure 2). In each case a new plasmid band was found with the same mobility as the donor plasmid. Southern hybridisation was performed with both pRK290 or pSUP106 as probes. The autoradiograms confirmed the presence of these plasmids in each transformant (Figure 2).

Besides the altered capability to take up DNA, either by conjugation or by transformation, we examined this mutant for other properties of this mutant (e.g. morphology, antibiotic resistance, substrate utilization). In all these tests (data not shown) no difference was found between A. brasilense 29710 and HE1, except that motility of the cells and their nitrogen fixation capability were reduced.

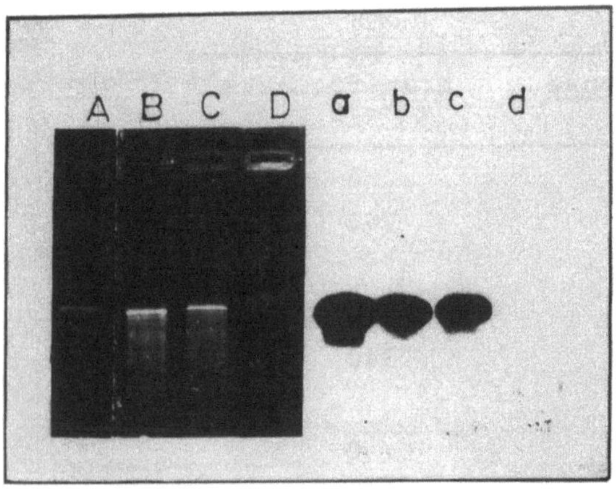

Fig. 2a) Evidence for transformation of A. brasilense HE1 with plasmid pRK 290.

Left: Agarose-gel. Right: Southern-blot.

B + C) = A. brasilense HE1 (pRK 290)

A = E. coli (pRK 290)

D = A. brasilense HE1 (without plasmid)

a,b,c,d = equal to A, B, C, D after hybridization with labelled pRK290

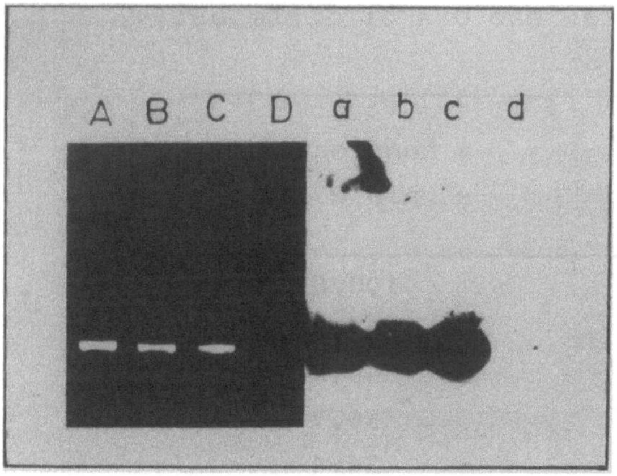

Fig. 2b) Evidence for transformation of A. brasilense HE1 with plasmid pSUP106 .

Left: Agarose-gel. Right: Southern-blot.

B + C) = A. brasilense HE1 (pSUP106)

A = E. coli (pSUP106)

D = A. brasilense HE1 (without plasmid)

a,b,c,d = equal to A, B, C, D after hybridization with labelled pSUP106

To confirm the identity of HE1 on the molecular level, further tests were made. DNA-homology studies were carried out using A. brasilense 29710 DNA as probe against DNAs of A. brasilense 29710 as control (100% homology), A. brasilense 29711, 29145 and HE1, A. lipoferum 29708 and 29731, Enterobacter cloacae and E. coli (Table 5). The results document complete homology between A. brasilense 29710 and HE1.

Tab. 5 Percent homology between the DNA of different Azospirillum strains

strain	% homology with strain 29710
A.brasilense ATCC29710	100.0
A.brasilense HE1	97.0
A.brasilense ATCC29711	69.4
A.brasilense ATCC29145	80.7
A.lipoferum ATCC29708	24.4
A.lipoferum ATCC29731	36.1
Ent.cloacae MF 10	10.0
E.coli HB 101	18.9

The plasmid patterns of both A. brasilense 29710 and HE1 were compared using the method of Eckhardt (10). The patterns are identical in both strains (Fig. 3).

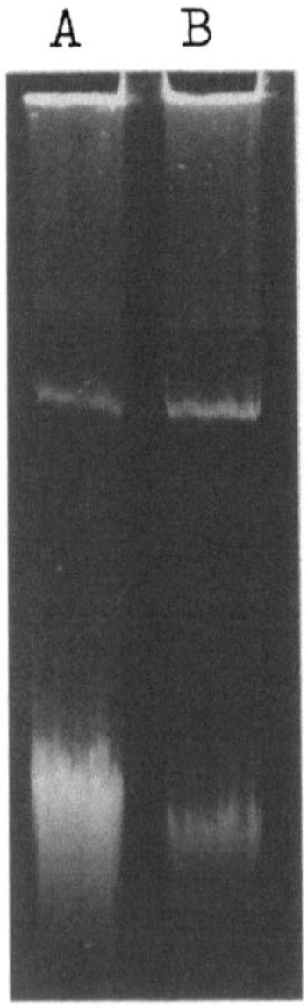

Fig. 3
Comparison of the plasmid pattern of the transformable (lane B) and the original strain (lane A).

The molecular basis of the mutation is still under investigation. The availability of strain HE1 hopefully will help to advance genetics of Azospirillum.

Acknowledgement

We thank Dr. W. Schilf, Dr. M.S. Abd-el-Salam and Dr. G. Schwabe for helpful discussions. Thanks are also due to B. Gubitz and G. Siebeker for typing the manuscript. This work was supported by the German Academic Exchange Service.

References

1. El-Khawas, H.M. (1981). M.Sc. Thesis. Fac. Agric., Cairo Univ.
2. Hegazi, N.A. (1983). In Klingmüller, W. (ed.) Azospirillum II: Genetics, Physiology, Ecology. Experientia Supplementum 48, Birkhäuser-Verlag, Basel, p. 171-189
3. Vlassak, K. and Reynders, L. (1978). Proc. Steenbock Kettering Int. Symp. N_2-Fixation, Abst. 44, Madison, U.S.A.
4. Fani, R., Bazzicalupo, M., Coianiz, P. and Polsinelli, M. (1986). FEMS Microb. Letters 35, 23-27
5. Schwabe, G. and Weber, U. (1983). In Klingmüller, W. (ed.) Azospirillum II: Genetics, Physiology, Ecology. Experientia Supplementum 48, Birkhäuser-Verlag, Basel, p. 56-65
6. Hanahan, D. (1983). J. Mol. Biol. 166, 557-580
7. Rodriguez, R. and Tait, R. (1983). Recombinant DNA techniques, an introduction. Addison-Wesley publishing company, London
8. Humphreys, G.O., Willshaw, G.A. and Anderson, E.S. (1975). Biochim. Biophys. Acta 383, 457-463
9. Chesney, R.H., Scott, J.R. and Vapnek, D. (1979). J. Mol. Biol. 130, 161-173

10. Eckhardt, T. (1978). Plasmid 1, 584-588

11. Ries, G., Holloway, B.W. and Pühler, A. (1980). Genet. Res. 36, 99-109

12. Maniatis, T., Fritsch, E.F. and Sambrook, J. (1982). Molecular cloning. Cold Spring Harbor Laboratory

13. Johnson, J.L. and Francis, B.S. (1975). J. General Microbiol. 88, 229-244

14. Simon, R., Priefer, U. and Pühler, A. (1983). Bio/technology 1, 784-791

15. Singh, M. and Klingmüller, W. (1986). Mol. Gen. Genet. 202, 136-142

16. Boyer, H.W. and Roullard-Dussoix, D. (1969). J. Mol. Biol. 41, 459-472

17. Dixon, R., Cannon, F.C. and Kondorosi, A. (1976). Nature 260, 268-271

18. Datta, N., Hedges, R.W., Shaw, E.J., Sykes, R.B. and Richmond, M.H. (1971). J. Bacteriol. 108, 1244-1249

19. Kleeberger, A. and Klingmüller, W. (1980). Mol. Gen. Genet. 180, 621-627.

20. Ditta, G., Stanfield, S., Corbin, D. and Helinski, D.R. (1980). Proc. Natl. Acad. Sci. (USA) 76, 1648-1652

21. Bagdasarian, M., Bagdasarian, M.M., Coleman, S. and Timmis K.N. (1979). In: Timmis, K.N. and Pühler, A., (eds.) Plasmids of medical, environmental and Commercial importance. Elsevier-North Holland, Amsterdam, p. 411-422

22. Bagdasarian, M., Franklin, F.C.H., Lurz, R., Rückert, B., Bagdasarian, M.M. and Timmis, K.N. (1981). Gene 16, 237-247

SELECTION AT THE CHEMOSTAT OF *Azospirillum brasilense* Cd N_2-FIXING AT HIGH O_2 PRESSURE

Maddalena Del Gallo, Loretta Gratani and Giorgio Morpurgo
Department of Plant Biology, University of Rome,
Piazzale A. Moro 5, I-00185 Rome, Italy

Summary: *A. brasilense* Cd was grown in a chemostat for 66 days, 32 change volumes, D=0.02 at atmospherical oxygen pressure. The selected strain was compared with the wild type at different O_2 levels steady states. Keywords: *Azospirillum*, O_2-tolerance, nitrogen fixation, chemostat continuous cultures.

Introduction

A. brasilense is a nitrogen-fixing bacterium living in the soil in association with the root system of various plants and, in particular, with Graminaceae (1).

Azospirilla are strictly aerobic microorganisms, but efficient nitrogen fixation occurs only at oxygen tensions of 0.01 atm or lower (2, 3). Higher O_2 partial pressure levels inhibit nitrogenase function inactivating the enzyme (4).

We describe here an attempt to select, through the chemostat, mutants of *A. brasilense* Cd insensitive to the O_2 repression of the nitrogenase. The rationale of the method was to cultivate in chemostat very large populations of this bacterium in a synthetic medium totally lacking combined nitrogen at O_2 tensions severely hampering nitrogen fixation. In these conditions a putative mutant insensitive to oxygen should substitute the wild type

Azospirillum IV: Genetics, Physiology, Ecology
Edited by W. Klingmüller

population. The selection of mutants through the chemostat is actually a useful method already described by Novick & Szilard (5) and recently reviewed by Dykhuizen & Hartl (6).

Material and Methods

A. brasilense Cd, ATCC 29729, isolated from roots of Cynodon dactylon (7), was grown at 30°C in a minimal medium (8) containing 33 mM malic acid neutralized with NaOH. The cultures were grown continuously in chemostats with working volumes of 1000 ml. The cultures were stirred magnetically (250 rpm) and were flushed through a gas sparger, with mixtures of N_2 and air whose composition was varied (0.21, 0.10, 0.05, 0.02, 0.01 atm O_2 pp) and controlled before entering the vessel by a Rank Brother Oxygen Electrode. The dilution rate was kept constant for all experiments at 0.02 with a double channel LKB 2132 MicroPerpex Pump. The culture vessel was inoculated and batch growth allowed to proceed for 24 h with a starter of yeast extract (0.1 g x l^{-1}) and $(NH_4)_2SO_4$ (0.2 g x l^{-1}) before the nutrient pump was started. 32 culture volumes were allowed to pass through the vessel during the first experiment of selection, 6 to 8 culture volumes in the second experiment before steady states were reached.

Optical density (OD) was measured spectrophotometrically at 545 nm. Cells were counted in an Improved Neubauer counting chamber. Nitrogenase activity was measured gas-chromatographically. The rate of oxygen uptake was measured in an oxygen electrode 3 ml volume cell (Rank Brother) at 30°C. Samples were stirred magnetically. Electrode was calibrated from zero to air saturation. The poly-β-hydroxybutyrate was measured by the Ward & Dawes disk method (9) using PHB dissolved in H_2SO_4 as a standard. Protein content was measured by the method of Lowry and al. (10) using bovine serum albumine as standard. Bacterial dry weight at the steady states was determined in triplicate by centrifuging and resuspending in distilled water 10 ml of culture samples and drying the cells to constant weight at 90°C. The supernatant obtained with the centrifugation was utilized for the analysis of the residual malic acid which was measured enzymatically with a Boehringer-Mannheim UV-Test.

Results

In a first experiment an attempt to cultivate A. brasilense Cd in synthetic nitrogen-free medium at atmospherical oxygen pressure (0.21 atm) was made (figure 1).

The selection was run for about 66 days at D 0.02 - about

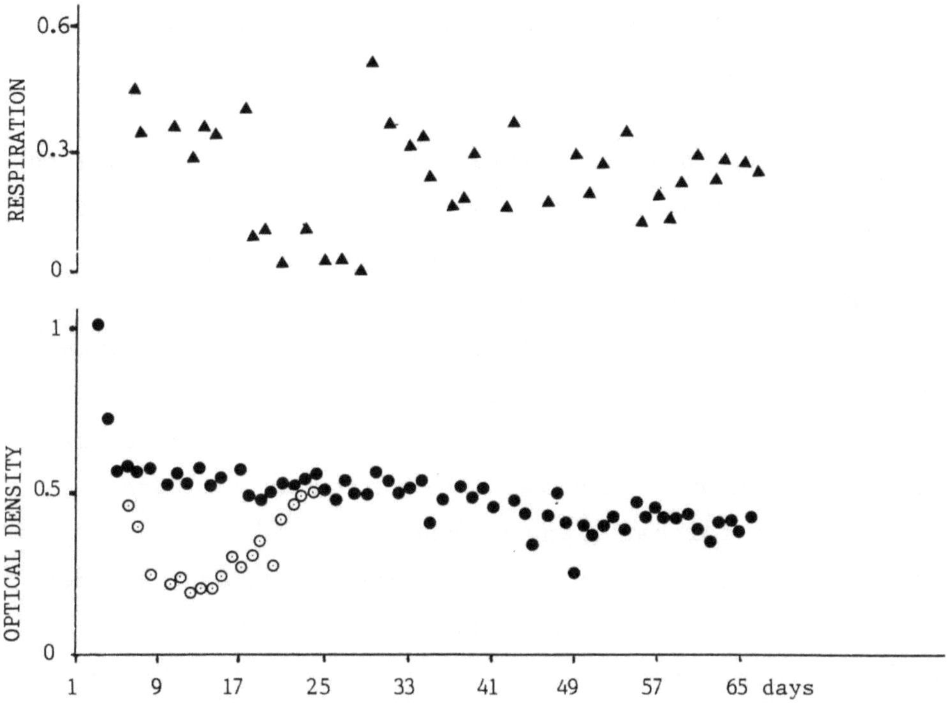

Figure 1. Selection at the chemostat of an A. brasilense Cd strain resistant to O_2 pp of 0.21 atm in nitrogen fixation condition. ●—●, optical density after vigorous shaking and after sedimentation of the clumps ○—○ . Respiration ▲—▲ (ug of O_2 uptaked by 1 ml of culture in 5 min).

32 volume changes. OD and respiration were checked daily. Because of cell clumping during the first days, OD of both samples were controlled. Surprisingly this bacterium may multiply indefinitely and fix nitrogen also at this oxygen level. As it can be noted from the figure, the OD of the population remained almost constant. This apparently means that mutants with high nitrogen fixation efficiency were not selected in the chemostat. But during this prolonged incubation at least three types of evident changes occurred: i) the clumps slowly reduced after the 15th day and desappeared completely around the 24th, ii) after the 17th day suddenly the culture turned from deep red to white, iii) the respiration dropped drastically at the same time from 0.5 mg x ml^{-1} x 5 min to 0.2 for 13 days, returning after that period of time - although irregularly - to higher levels.

The clumps are probably an adaptative phenomenon to reduce the effect of high oxygen levels; they slow down the diffusion of the oxygen at the inner cells, which can therefore fix nitrogen. The red color, due to carotenoids syntesized in nitrogen fixation condition according to Nur et al. (11), provide an efficient protection from oxygen. The colour of this mutant remained permanently white in liquid culture, but returned red when streaked on the surface of solid medium. The selection was interrupted at the 66th day because no evident further changes occurred.

A second experiment with this newly obtained mutant was performed. The oxygen level was gradually reduced from 0.21 to 0.10, 0.05 and 0.01 atm, and some physiological relevant parameters were recorded at each different steady state: i.e. C_2H_2 reduction, number of bacteria, OD, respiration, protein content, PHB % of cells and residual malate in the culture medium. The data are shown in figure 1, 2 and table 1.

At the same time a second chemostat was started. Nutritional

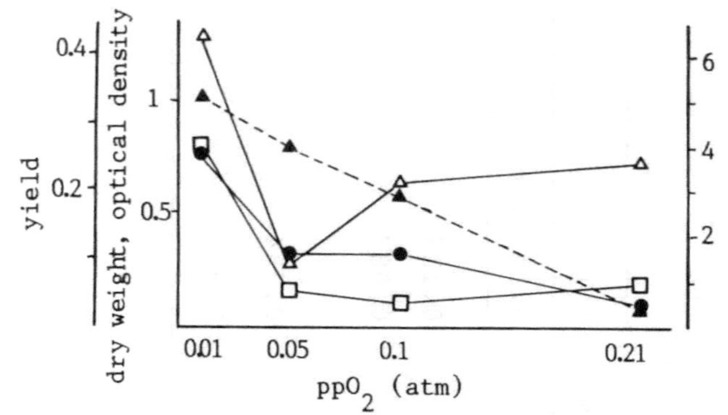

Figure 2 (mutant). Optical density □——□. Respiration mg O_2/proteins x min △——△ .Dry weight mg x ml ●——●. Growth yield (Y), mg dry weight/mg malate consumed x ml ▲----▲ .

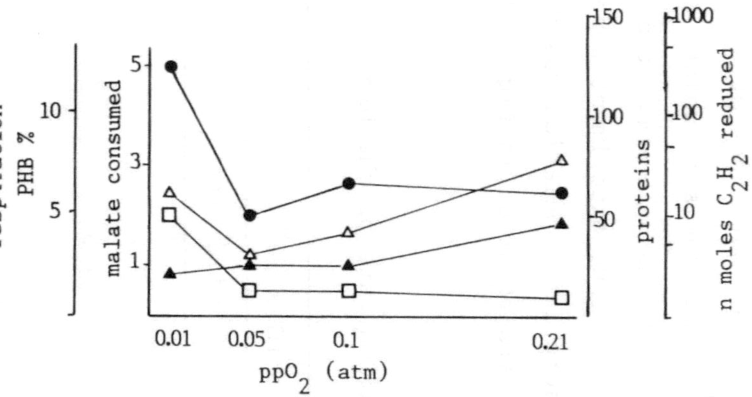

Figure 3 (mutant). Proteins □——□ , mg x ml x 10^{-3}. Malate consumed, mg x ml △——△ . PHB, % dry weight ▲——▲ . n moles of C_2H_2 reduced / mg proteins x h ●——● .

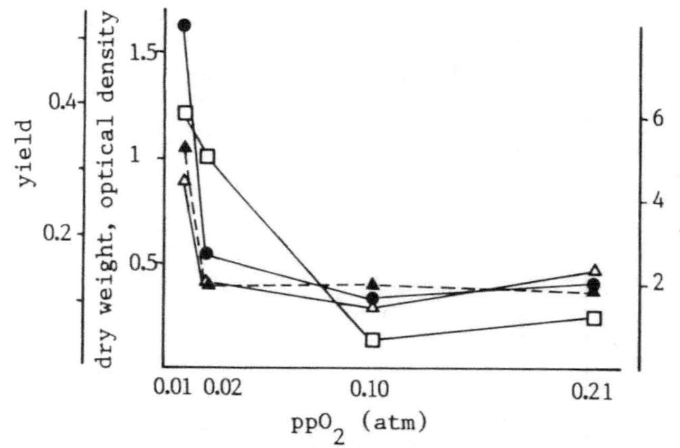

Figure 4 (wild type). Symbols as in fig. 2

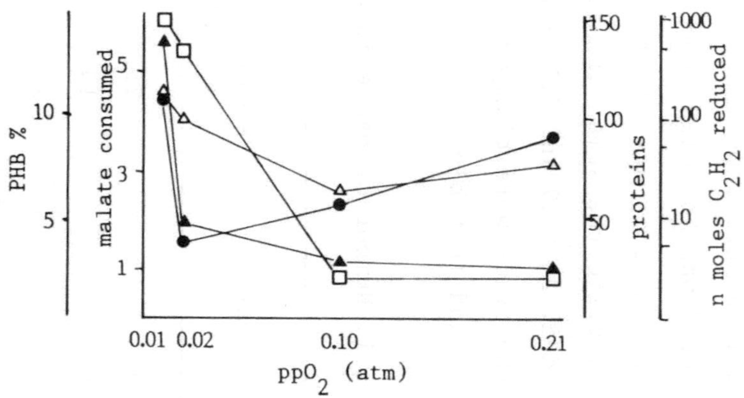

Figure 5 (wild type). Symbols as in figure 3

conditions were identical to the previous one, but the strain incubated was the wild type Cd and the oxygen levels to which it was exposed increased from 0.01 to 0.02, 0.10 and 0.21 atm. At each steady state the same parameters of the previous experiment were recorded. At 0.10 and 0.21 atm O_2 the bacterial population was red and the cells were clumped. The results are reported in figure 4 and 5, and in table 1.

Table 1. Number of bacteria and Respiration rate of the biomass

pO_2 (atm)	N. of bacteria x ml x 10^6	
	mutant	wild type
0.21	25.5	7.2
0.10	19	14.5
0.05	17	-
0.02	-	170
0.01	74	216

pO_2 (atm)	mg O_2 / dry weight (x ml)	
	mutant	wild type
0.21	0.580	0.117
0.10	0.055	0.087
0.05	0.125	-
0.02	-	0.534
0.01	0.230	0.408

Discussion

If we compare the behaviour of the mutant with that of the wild type, at first glance we could conclude that A. brasilense cannot be genetically improved toward environmental conditions in which it is necessary to fix nitrogen at atmospherical O_2 pressure. Surprisingly the wild type strain gave better results than the putative resistant mutant in i) C_2H_2 reduction, ii) dry weight, iii) protein content, and it seems therefore more adapted. However, the number of cells per ml at high O_2 partial pressure is more than three times higher in the mutant than in the wild type and consequently the cells are smaller. This means that the fitness of the mutant is - as expected - much higher than that of the wild type. It must also be noted that the PHB content of the mutant increases at the highest O_2 pressure, while in the wild type it decreases. PHB may in fact offer some protection for the nitrogenase toward atmospheric oxygen pressure (12).

The respiration of the mutant is also strongly increased either when measured as mg O_2 consumed per ml of culture or per dry weight. In this case respiration is five-fold higher than that of the wild type. At lower oxygen tensions respiration is, on the contrary, only one-half that of the wild type.

In conclusion the experiments demonstrate that A. brasilense may increase its fitness in conditions of nitrogen deprivation at atmospherical O_2 pressure, also if its productivity - measured as OD, N_2-fixation or dry weight - does not increase.

The mutant obtained shows profound physiological differences in comparison with the wild type, but it is difficult to understand, at the moment, which of these alterations are related to the high fitness of the mutant.

The fact that in conditions of high oxygenation it does not

form clumps and does not synthesize carotenoids suggests that its N_2-fixation system is, in some way, more protected and/or resistant to oxygen. In fact the single cells of the mutant are really exposed to high oxygen pressure while the wild type very likely may grow only because of the reduced oxygen tension inside the clumps.

Further studies are necessary however to understand the physiological bases of this phenomenon.

References

1. Okon, Y. (1985). Trends Biotechnol, 2, 223-228.
2. Nelson, L.M., Knowles, R. (1978). Can. J. Microbiol. 24, 1395-1403.
3. Fu, C., Knowles, R. (1986). Can. J. Microbiol. 32, 897-900.
4. Ludden, P.W., Okon, Y., Burris, R.H. (1978). Biochem. J. 173, 1001-1003.
5. Novick, A., Szilard, L. (1950). Proc.N.A.S. USA 36, 708-719.
6. Dykhuizen, D.E., Hartl, D.L. (1983). Microbiol. Rev. 47, 150-168.
7. Eskew, D.L., Focht, D.D., Ting, I.P. (1977). Can. J. Microbiol. 34, 582-585.
8. Martinez-Drets, G., Del Gallo, M., Burpee, C., Burris, R.H.(1984) J. Bacteriol. 154, 80-85.
9. Ward, A.C., Dawes, E.A. (1973). Anal. Biochem. 52, 607-613.
10. Lowry, O.H., Rosenbrough, N.J., Farr, A.L., Randall, R.J. (1951). J. Biol. Chem. 193, 265-275.
11. Nur, I., Okon, Y., Henis, Y. (1982). Isr. J. Bot. 31, 221-227.
12. Tal, S. Okon, Y. (1987). Can. J. Microbiol. 31, 608-613.

ROOT HAIR DEFORMATION INDUCED ON MAIZE AND MEDICAGO BY AN AZOSPIRILLUM TRANSCONJUGANT CONTAINING A RHIZOBIUM MELILOTI NODULATION REGION

L. Piana, M. Delledonne, M.N. Antonelli, C. Fogher

Istituto di Genetica agraria, Università Cattolica,
via E. Parmense 84, 29100 Piacenza, Italy

Summary: Rhizobium bacteria form nitrogen-fixing root nodules on legume plants. This process requires the induction and expression of genes in both partners. After cloning of the nod hsn region from the pSym of R. meliloti we introduced this region in A. brasilense. Using the plate method we analysed the phenotype induced on root hairs of maize and alfalfa by this transconjugant. When plasmid pPF1 was mobilized into A. brasilense AB11 the transconjugant caused a significant increase in root hair deformation, in comparison with the wild type, both on maize and alfalfa. The meaning of the nod genes expression in the heterologous bacterial host is discussed.

Keywords: Azospirillum transconjugant, root hairs, nodulation genes.

Introduction

The bacteria of the genus Rhizobium have the ability to establish nitrogen-fixing associations with their host plants (1). This process requires the recognition and the infection of root hairs followed by the development and differentiation of root nodules (2). In Azospirillum no gene involved in bacteria-plant interaction has yet been identified (3). Recently homology of Azospirillum DNA with nod and hsn probes of Rhizobium was found (4). As Azospirillum contains large plasmids (5), it is possible that some of them are related to plasmids found in Rhizobium, consequently Rhizobium symbiotic genes could be expressed in Azospirillum, inducing some of the early steps of the recognition between bacteria and plant (e.g. root hair curling). In this paper we describe the phenotype induced on

Azospirillum IV: Genetics, Physiology, Ecology
Edited by W. Klingmüller

plants by an Azospirillum strain containing the symbiosis region of R. meliloti.

Materials and methods

Bacterial strains and plasmids

LB for E. coli have been described (6). Media for Azospirillum were complete medium D (7) and minimal medium K (5). Concentration of antibiotics used for Azospirillum were 10 µg of tetracycline and 40 µg of kanamycin per ml. A. brasilense AB11 (SPF13) is a his^- Rif^r Sp6 (8) derivative obtained by Bazzicalupo (Florence). E. coli strains were HB101 (pro, leu, thi, lacY, endoI, recA) and LE392 (hsd, met, trp, supE, supF). Plasmid pPF1 is a pVK100 (9) derivative containing about 30 kb of the pSym of R. meliloti strain 2011 (10). Plasmid pRK2013 was in HB101 and used as helper (Tra^+) in tri-parental matings (11).

Genetic techniques

Bacterial mating was performed using standard techniques. Donor and recipient cultures were grown in a rich medium at a concentration of about 5×10^8 bacteria per ml. They were mixed in the proportion of 1:1:8, donor: helper: recipient. Conjugation mixture was spread and incubated overnight on a rich medium. The bacteria on the plate were then suspended in 2 ml of minimal medium and spread on selective plates and incubated at 30° C. The cosmid clone LE392(pPF1) was selected after colony hybridization of a bank of the plasmid pSym2011 (Piana L., Fogher C., unpublished), with probes from pKSK5 and pEK10 (12). Plasmid pPF1 gave a positive response with both probes and was presumed to contain the common nod and the hsn regions of pSym2011 (12, 13). The purified plasmid was transformed in HB101 and used in the conjugation.

Plant material

Maize (Zea mays L., BC_2S_1 from Mexican germplasm crossed with the inbred line Mo17) and alfalfa (Medicago sativa L., cv Bresaola) seeds were sterilized in 70% ethanol for two min and in 20% sodium hypochlorite for 15 min, and washed four times with sterile water. Seedlings for assay were germinated on sterile water agar plates and, when the root was about one cm long, transferred to square plates containing nitrogen-free Fahraeus medium (14, 15). The inocula were grown for 24 h in liquid medium and then used to inoculate the plants, with 20 µl of culture per plant, distribuited over about 2-3 cm of root length. Inoculated plates were incubated vertically, with the root portion covered with black plastic film, in a growth chamber at 25° C and continous light of 5,000 mW/m^2. The root hair morphology was observed at 5, 10 and 30 days using an inverted microscope. All root hairs projecting laterally from the root were examined for root hair deformations.

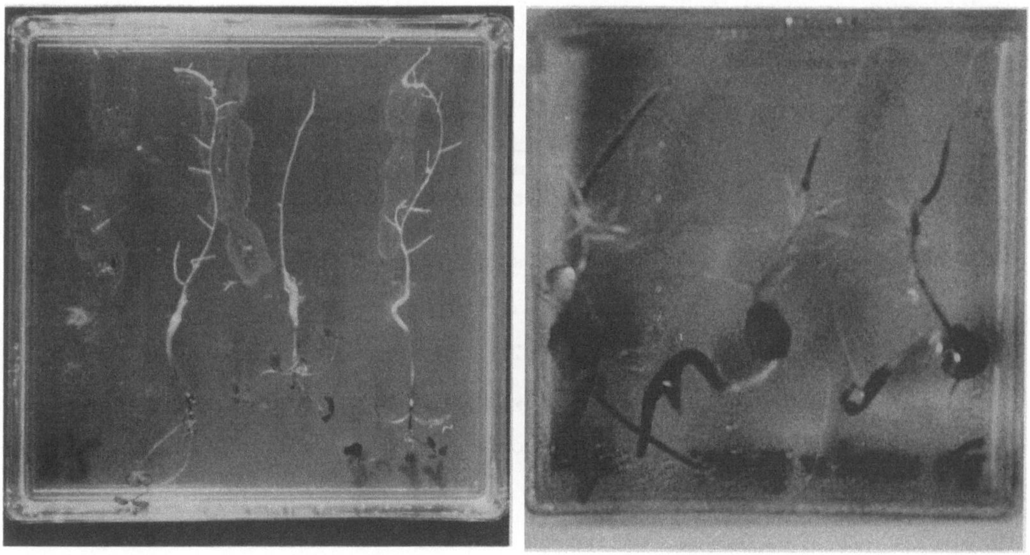

Figure 1. Inoculated seedlings of maize (left) and alfalfa (right) on plate.

Results

To study the interaction of various rhizobial strains, the plate method was succesfully used (14, 16), and the incubation of inoculated seedlings on agar plates (figure 1) provided a procedure for rapid and continuous control of the phenotype induced on root hairs. When plasmid pPF1 was mobilized into A. brasilense strain AB11, the transconjugant (AB11nod) gave a significant increase in root hair deformation compared with the wild type, both on maize and medicago (table 1).

Table 1. Root hair deformation of tested plants after inoculation with different strains (%).

Strain	Alfalfa			Maize		
	days after inoculation					
	5	10	30	5	10	30
2011	20	30	90	0	0	0
AB11	0	0	0.5	0	1.5	3
AB11nod	1.5	8	40	0	2	20

Several microscopic fields were examined per plant and a minimum of 20 plants were examined per inoculated strain.

The examination of the infected plants showed that the ability to curl alfalfa root hairs (Hac^+ phenotype) was morphologically comparable in R. meliloti 2011 and AB11nod (figure 2). With maize the root hair deformation was less charatteristic and, in general, we considered as positive response to bacterial infection any root hair that had grown through a 120° turn (figure 3).

Discussion

Rhizobium bacteria form nitrogen-fixing root nodules on legume plants. This process requires the induction and expression of genes in both part-

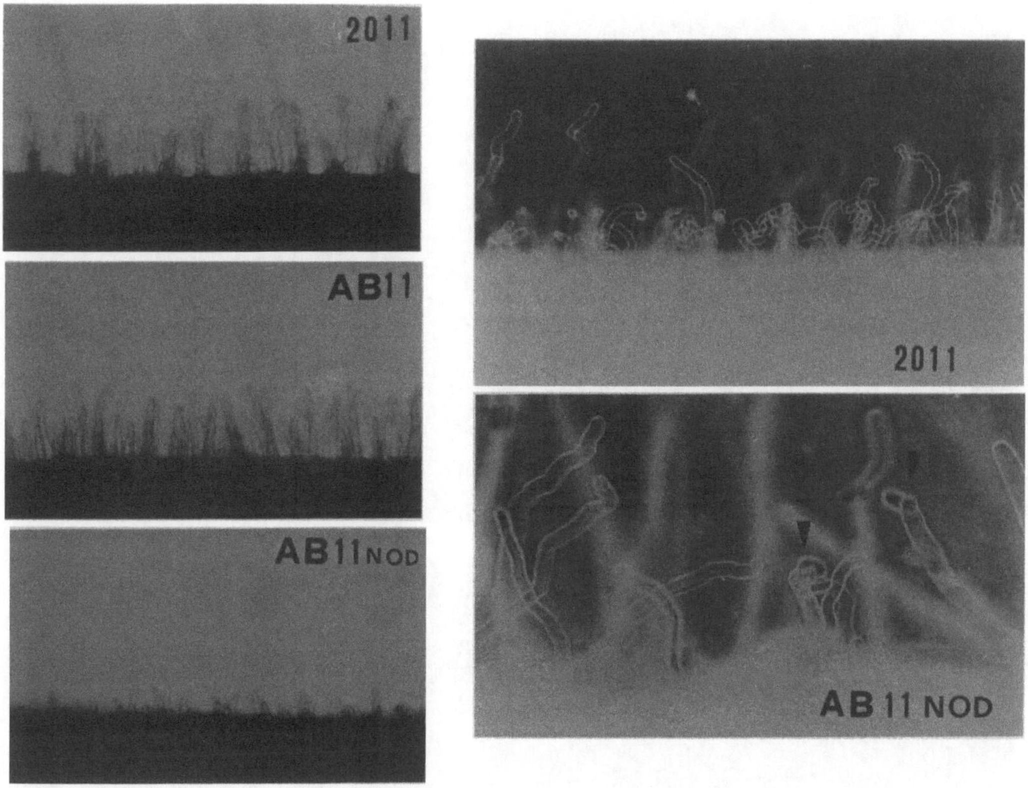

Figure 2. Root hairs of Medicago sativa infected by R. meliloti 2011, A. brasilense AB11 and A. brasilense AB11nod. Shepherd's crooks are indicated by arrows.

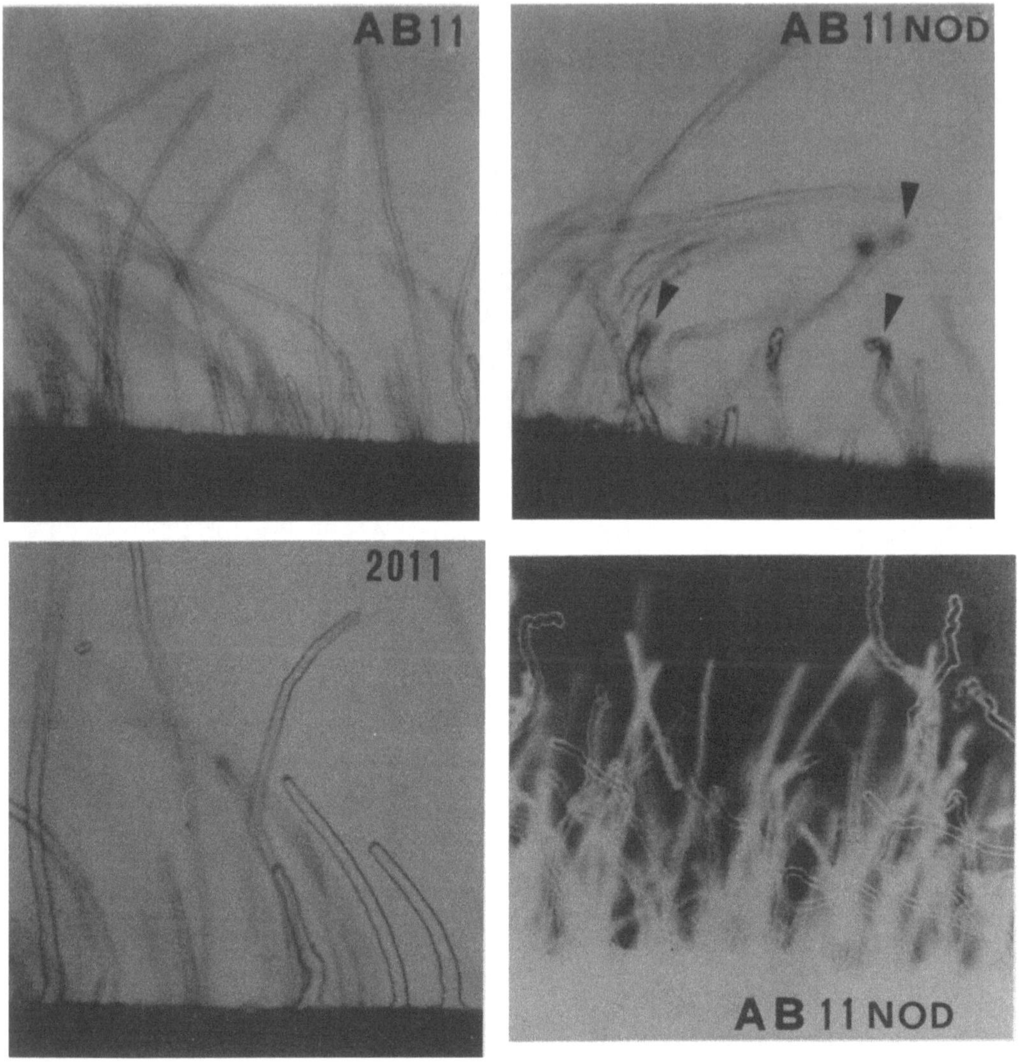

Figure 3. Root hairs of Zea mays infected with R. meliloti 2011, A. brasilense AB11 and A. brasilense AB11nod. Root hair deformation are indicated by arrows.

ners (17). The process of recognition, infection and nodule development is complex and requires several closely linked genes (1, 2). In _R. meliloti_ genes involved in attachment, root hair curling and nodule organogenesis are on a megaplasmid (12,18). These genes are clustered in two regions; the first group, the common _nod_ genes, is located on a fragment situated 6 kb away from the host-specificity nodulation (_hsn_) region (12). Using internal sequence of these two regions as probes, we have identified the plasmid pPF1 that, with an insert of 30 kb, should contain all the genes required for the first steps of the bacterial interaction with plants. This seems to be the case since root hair curling and pseudonodule formation (data not shown) occur on alfalfa infected with AB11nod. The other implication of this result is that the _nod_ genes are functional in the heterologous bacterial host. Hooykaas _et al_. (19) first used _A. tumefaciens_ as a recipient for _Rhizobium_ plasmids and the result was the formation of nodule-like structures on the pSym specific host plant. Examples of _nod_ gene transfer and expression in bacteria not belonging to the _Rhizobiaceae_ family is reported by Plazinski and Rolfe (20) and Hirsch _et al_. (21). In the first case the recipient was a _Lignobacter_ strain and there was a copy number effect on the host range expression of the _nod_ genes cloned. A high gene expression (multicopy) seems to be sufficient to obtain a root hair curling phenotype on monocotyledonous plants (22), at least using an homologous bacterial host.

This is the first report of symbiotic gene expression on a monocotyledonous plant with an heterologous (referring to the symbiosis region) bacterial host. The high copy number effect on maize, like that observed by Plazinski _et al_. (22), using a low number vector (pVK100 is a pRK290 derivative) could be explained by the presence of _nod_ and _hsn_ homologous regions on the genome of the heterologous _Azospirillum_ host.

Acknowledgements

M.N.A. was on leave from Enichem Agricoltura; L.P. was the recipient of a postdoctoral fellowship from research contract funds. We are grateful to Dr. A. Kondorosi for the gift of plasmids carrying the probes used.

References

1. Vincent, J.M. (1980). In W.E. Newton and W.H. Orme-Johnston (ed.), Nitrogen Fixation. University Park Press, Baltimore
2. Nuti, M.P., Lepidi, A.A., Prakash, R.K., Hooykaas, P.J.J. and Schilperoort, R.A. (1982). In Molecular Biology of Plant tumors; Kahl, G. and Schell, J.S. (eds.). Acad. Press, New York
3. Elmerich, C. (1984). Biotechnology 2, 967-978
4. Fogher, C., Dusha, I., Barbot, P. and Elmerich, C. (1985). FEMS Microb. Letters 30,245-249
5. Franche, C. and Elmerich, C. (1981). Ann. Microbiol. (Inst. Pasteur) 132A,3-17
6. Bozouklian, H., Fogher, C. and Elmerich, C. (1986). Ann. Microbiol. (Inst. Pasteur) 137B;3-18
7. Gauthier, D. and Elmerich, C. (1977). FEMS Microbiol. Letters, 2;101-104.
8. Bani, D., Barberio, C., Bazzicalupo, M., Favilli, F., Gallori, E. and Polsinelli, M. (1980). J. Gen. Microbiol. 119,239-244
9. Knauf, V.C. and Nester, E.W. (1982). Plasmid 8,45-54
10. Rosenberg, C., Boistard, P., Dénarié, J. and Casse-Delbart, F. (1981) Mol. Gen. Genet. 184,326-333
11. Ditta, G., Stanfield, S., Corbin, D. and Helinski, D. (1980). Proc. Natl. Acad. Sci. USA 77,7347-7351
12. Kondorosi, E., Banfalvi, Z. and Kondorosi, A. (1984). Mol. Gen. Genet. 193,445-452.
13. Julliot, J.S., Dusha, I., Renalier, M.E., Terzaghi, B., Garnerone, A.M. and Boistard, P. (1984). Mol. Gen. Genet. 193,17-26

14. Rolfe, B.G., Gresshoff, P.M. and Shine, J. (1980). Plant Sci. Letters 19,277-284

15. Vincent, J.M. (1970). Int. Biology Prog. Handbook No. 15, Blackwell, oxford

16. Djordjevic, M.A., Zurkowski, W., Shine, J. and Rolfe, B.G. (1983). J. Bacteriol. 156,1035-1045

17. Verma, D.P.S. and Long, S.R. (1983). Int. Rev. Cytol. (Suppl.)14,211-245.

18. Banfalvi, Z., Sakanyan, V., Koncz, C., Kiss, A., Dusha, I. and Kondorosi A. (1981). Mol. Gen. Genet. 184,318-325

19. Hooykaas, P.J.J., van Brussel, A.A.N., de Dulk-Ras, H., van Slogteren, G.M.S. and Schilperoort, R.A. (1981). Nature 291,351-353

20. Plazinski, J. and Rolfe, B.G. (1985). J. Bacteriol. 162,1261-1269

21. Hirsch, A.M., Wilson, K.J., Jones, J.D.G., Bang, M., Walker, V.V. and Ausubel, F.M. (1984). J. Bacteriol. 158,1133-1143

22. Plazinski, J., Innes, R.W. and Rolfe, B.G. (1985). J. Bacteriol. 163,812-815

INFLUENCE OF AZOSPIRILLUM BRASILENSE ON NODULATION OF LEGUMES

W. Schmidt*, P. Martin*, S.H. Omay*, F. Bangerth**
*Institut für Pflanzenernährung, **Institut für Obst-, Gemüse- und Weinbau, Universität Hohenheim, Postfach 7oo562, D-7000 Stuttgart 70, FRG

Summary: Combined inoculation of *Rhizobium meliloti* and *Azospirillum brasilense* Cd to seedlings of *Medicago sativa* grown in large petri dishes on sterile agar increased the number of nodules significantly but the effects depended on the concentrations of *Azospirillum* in the medium. The stimulating effect is probably due to the production of indole acetic acid (IAA) by *Azospirillum* since pure IAA mixed to the agar also led to higher numbers of nodules. An increase in nodule number was also found with soybean grown in soil under nonsterile conditions upon repeated inoculation with *Azospirillum*. The strain used and other *A.brasilense* strains produced considerable amounts of IAA when cultivated in N-containing nutrient solution whereas release from *A. lipoferum* strains was very low. A time course showed that most production of IAA occurred after the end of the logarithmic growth phase. The results indicate that rhizosphere bacteria may play a role in nodulation of legumes through hormonal effects. Keywords: Nodulation, indole acetic acid (IAA), legumes, *Azospirillum*.

Introduction

Free living N_2-fixing bacteria in the rhizosphere may have a beneficial effect on plant growth by supplying fixed nitrogen, especially when the bacteria are enriched by inoculation. However, a number of experiments have shown that root morphology may also be affected upon inoculation. Increased root branching or higher root hair density has been observed after inoculation with *Azo-*

Azospirillum IV: Genetics, Physiology, Ecology
Edited by W. Klingmüller

spirillum (1, 4, 8, 9, 11, 18). It is still a matter of controversy of whether the stimulation of plant growth by inoculation with Azospirillum is due to nitrogen fixation or to an increase in root surface and the resulting increased uptake of nutrients, or both. A. brasilense may produce phytohormones, especially IAA, when cultured in nutrient solution (4, 6, 13, 18). Such hormones may be responsible for the observed changes in root morphology.

The newly observed increase in nodule number at roots of leguminous plants after combined inoculation of Rhizobium and Azospirillum (7, 15, 16, 17, 19) possibly is also due to the production of hormones by Azospirillum. Not very much is known of the role rhizosphere bacteria play in the root infection by Rhizobium and subsequent nodule formation. In the present paper A. brasilense was used as a model bacterium in experiments on nodulation of Medicago sativa and Glycine max. The effect of inoculation of Azospirillum brasilense was compared with the effect of the addition of pure IAA.

Materials and Methods

Seeds of Medicago sativa were surface disinfected (5 % NaOCl) and germinated under sterile conditions in petri dishes on agar. Seedlings were selected after 2 days and transferred to large petri dishes (diameter 200 mm), whose bottom was covered with a 5 mm layer of agar (1,5 % agar). The agar contained macro-nutrients except N (mM/l: 8 $CaSO_4$, 1 KH_2PO_4 , 1 $MgSO_4$, 2 K_2SO_4) and microelements in adequate amounts. The pH was adjusted to 6.0. Rhizobium meliloti (DSM 1981) was mixed with the agar (10^6 CFU/ml) before pouring on the plate when the agar was cooled down to 40°C. In the same way Azospirillum brasilense Cd (10^3 to 10^7 CFU/ml), precultivated for 18 h, was added. For controls corresponding concentrations of autoclaved Azospirillum were mixed to the agar. The petri dishes each containing 6 seedlings which have been fixed with paper strips were put upright in a constant environment (16 h day, 22°C, irradiance 60 W m^{-2}). The roots were protected against light by putting the petri dishes in pouches of black foil. Apart from the sterile conditions, this type of agar assay allowed a continuous control of the root development throughout the experiment (3 weeks).

Glycine max (Effi) was cultivated in a soil/vermiculite mixture (wt 70/30). Seedlings were inoculated at planting with Bradyrhizobium japonicum (10^7 CFU/plant) and A. brasilense Cd (10^7 CFU/plant). Inoculation with A. brasilense was repeated 4 times (every 2nd day).

For inoculation A. brasilense Cd was cultivated under continuous shaking in a nutrient solution containing malate (10). To increase the buffer capacity of the basic solution phosphate buffer was added as described by Okon et al. (12). For the investigation of hormone production by Azospirillum yeast extract in the medium was replaced by 1 g NH_4Cl/l. Before inoculating Rhizobium/Bradyrhizobium and Azospirillum the bacteria were washed twice with phosphate buffer (ph 6.8). IAA proved to be stable during autoclaving and therefore was added to the agar medium before sterilisation.

IAA, abscisic acid (ABA) and zeatin/zeatinriboside were determined by radioimmunoassay.

Vertical bars in Figures represent standard error of the mean.

Results and Discussion

Fig. 1 shows that the effect of A. brasilense Cd on nodulation depends on the number of CFU added to the medium. Lower concentrations of CFU stimulated the nodule number significantly

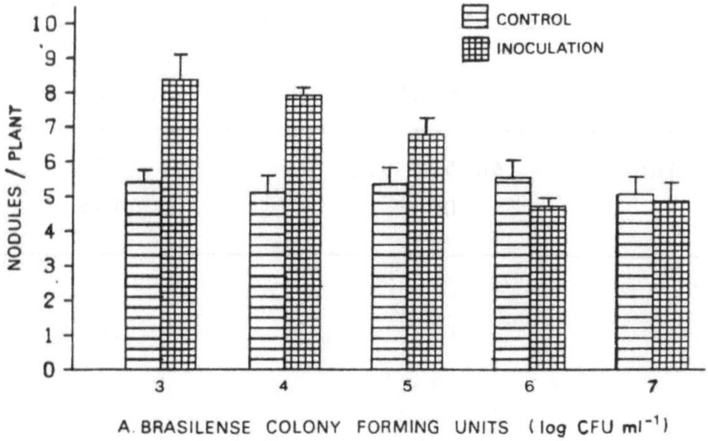

Fig. 1. Influence of *A. brasilense* Cd on nodulation of *Medicago sativa*.

whereas at higher concentrations nodules tended to decrease. Plazinski and Rolfe (14) also found that in such testing systems (agar assay) either an increase or decrease in nodulation can result from inoculation depending on the ratio *Rhizobium*/*Azospirillum* and on the time of inoculation of *Azospirillum* in relation to inoculation with *Rhizobium*. Root branching was decreased by inoculation with *Azospirillum* in the experiment whose results are given in Fig. 1 (data not shown). Therefore the increase in nodulation cannot be ascribed to a possible increase of suitable sites for infection by an increased number of lateral roots. The lack of stimulation of nodulation at higher concentrations of CFU is paralleled by an inhibition of root elongation (data not shown).

In accordance with results of Singh and Subba Rao (17) and Sarig et al. (16) *Azospirillum* also stimulated nodulation when plants were grown under nonsterile conditions. For the experiment whose results are given in Table 1 *Glycine max* was cultivated in a soil-vermiculite mixture.

To investigate the effect of IAA on nodulation *M.sativa* was cultivated in agar plates as before but instead of *Azospirillum*

Table 1. Influence of *A. brasilense* on nodulation of *Glycine max* (± SEM)

	control	inoculation
nodules/plant	66.5 ± 9.6	90.1 ± 5.7
%	100	135

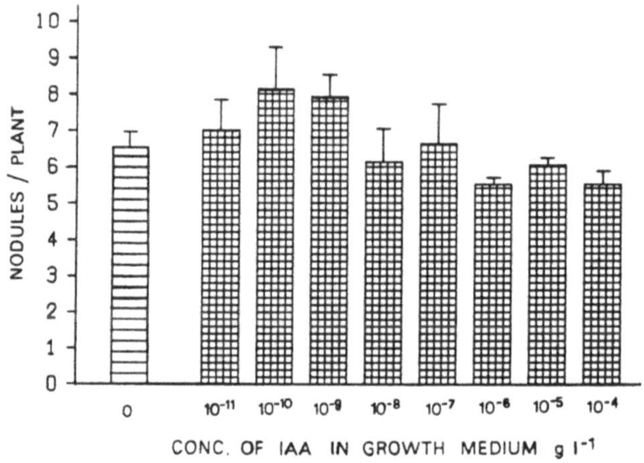

Fig. 2. Influence of IAA on nodulation of *Medicago sativa*.

different concentrations of pure IAA were added. Fig. 2 shows that the number of nodules was increased in the concentration range from 10^{-11} to 10^{-9} g/l. At higher concentrations nodulation was inhibited. Plazinski and Rolfe (15) investigating inhibitory effects on nodulation found depressions of nodulation by both *Azospirillum* and auxins. They suggest that at certain concentrations of IAA nodulation may be stimulated.

The corresponding stimulating effects of application of IAA and inoculation with *Azospirillum* on nodulation in the present experiments suggest that *Azospirillum* stimulates nodulation by the production of IAA. The hormone may exert its influence by a change in root morphology, e.g. of root hair shape or density.

On the other hand IAA could affect the mechanisms of the plant by which the number of nodules is regulated. Such regulating factors, which are not yet identified, may be produced in the root itself or transported from the shoot to the root (3).

Though it is repeatedly found that A. brasilense may produce IAA (2, 4, 6, 18) it was of interest to know whether the strain used is a weak or strong producer of IAA. Therefore the enrichment of IAA in the bacterial medium (NH_4-containing) of different strains was investigated. For comparison strains of A. lipoferum were included. Since there is evidence that Rhizobium/Bradyrhizobium also produces IAA some strains of these species were also tested. IAA determination was carried out after the end of the logarithmic phase. All strains of A. brasilense, including the strain used for the present experiments produced relatively high amounts of IAA, whereas in the medium of the A. lipoferum strains only traces of IAA were found (Fig. 3). One strain

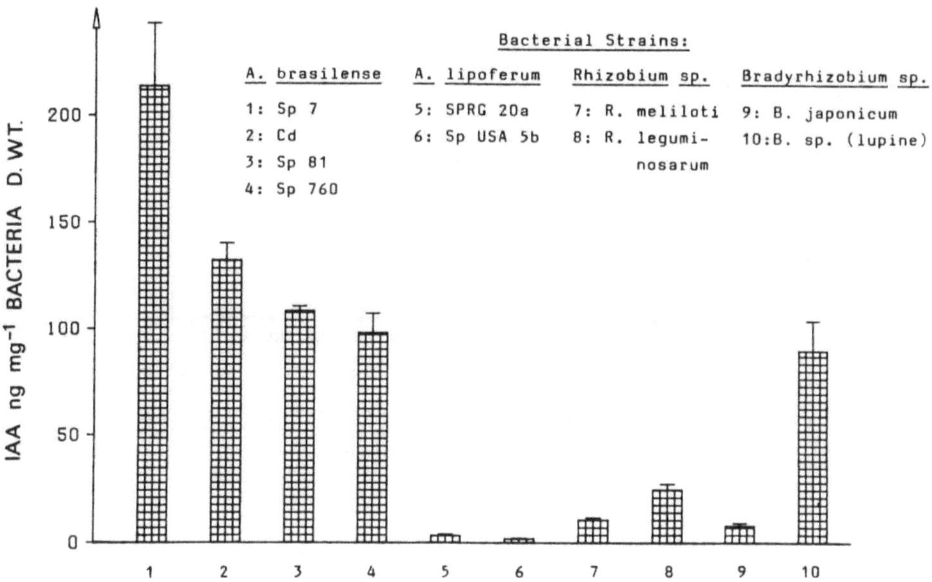

Fig. 3. IAA accumulated in the nutrient solution at the end of the logarithmic phase. Amounts detected are referred to bacterial dry weight.

of Bradyrhizobium produced high amounts similar to that of A. brasilense. IAA production of A. brasilense in a medium without added N remained very low (data not shown). ABA and zeatin/zeatinriboside were detected only in small amounts.

The time course of proliferation and IAA production of A. brasilense Cd shows that the main production of IAA starts only at the beginning of the stationary phase confirming results of Horemans and Vlassak (5) and De Francesco et al. (2) (Fig. 4).

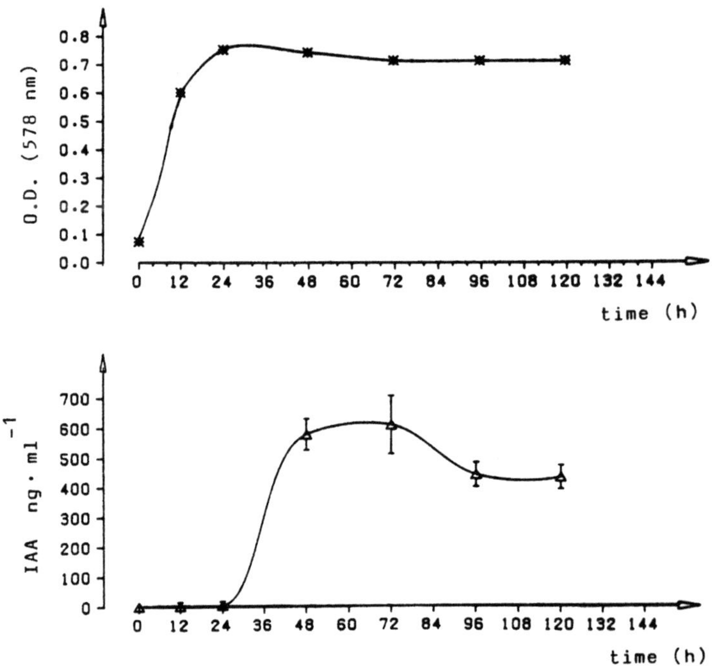

Fig. 4. Time course of optical density and IAA release from A.brasilense Cd into the nutrient medium.

Acknowledgement

This work has been supported by the Deutsche Forschungsgemeinschaft.

References

1. Barbieri, P., Zanelli, T., Galli, E., Zanetti, G. (1986). FEMS Microbiology Letters 36, 87-90

2. De Franceso, R., Zanetti, G., Barbieri, P., Galli, E. (1985). Azospirillum III: Genetics, Physiology, Ecology, (Ed. Klingmüller W.) pp. 109-115 Springer Verlag Berlin

3. Delves, A.C., Matthews, A., Day, D.A., Carter, A.S., Carroll, B.J. and Gresshoff, P.M. (1986). Plant Physiol. 82, 588-590

4. Hartmann, A., Fußeder, A., Klingmüller, W. (1983). In Azospirillum II, Genetics, Physiology, Ecology (Ed. Klingmüller, W.) Experientia Suppl. 48, 78-88

5. Horemans, S., Vlassak, K. (1985). Azospirillum III: Genetics, Physiology, Ecology, (Ed. Klingmüller, W.) pp. 98-108, Springer Verlag Berlin

6. Horemans, S., de Koninck, K., Neuray, J., Hermans, R., Vlassak, K. (1986). Symbiosis 2, 341-346

7. Iruthayathas, E.E., Gunasekaran, W., Vlassak, K. (1983). Scientia Horticulturae 20, 231-240

8. Kapulnik, Y., Gafny , R., Okon, Y. (1985). Can. J. Bot. 63, 627-631

9. Kolb, W., Martin, P. (1985). Azospirillum III, Genetics, Physiology, Ecology, (Ed. Klingmüller, W.) pp. 215-221 Springer Verlag Berlin

10. Krieg, N.R., Döbereiner J. (1984). Genus Azospirillum. In: Bergey's Manual of Systematic Bacteriology. 1, 94-104, N.R. Krieg and J.G. Holt (eds.) Williams and Wilkins, Baltimore

11. Martin, P., Glatzle, A. (1982). In Azospirillum, Genetics, Physiology, Ecology, (Ed. Klingmüller, W.) Experientia Suppl. 42, 108-120

12. Okon, Y., Houchins, J.P., Albrecht, S.L., Burris, R.H. (1977). J. Gen. Microbiol. 98, 87-93

13. Okon, Y., Kapulnik, Y. (1986). Plant Soil 90, 3-16

14. Plazinski, J., Rolfe, B.G. (1985). Appl. Environ. Microbiol. 49 (4), 984-989

15. Plazinski, J., Rolfe, B.G. (1985). Appl. Environ. Microbiol. 49 (4), 990-993

16. Sarig, S., Kapulnik, Y., Okon, Y. (1986). Plant Soil 90 (3), 335-342

17. Singh, C.S., Subba Rao, N.S. (1979). Plant Soil 53, 387-392

18. Tien, T.M., Gaskins, M.H., Hubbell, D.H. (1979). Appl. Environ. Microbiol. 37 (5), 1016-1024

19. Tilak, K.V.B.R., Singh, C.S., Rana, J.P.S. (1981). Zbl. Bakt. II.Abt. 136, 117-120

ARE AZOSPIRILLUM BACTERIOCINS PRODUCED AND ACTIVE IN SOIL?

R.G.B. Oliveira and A. Drozdowicz
Institute of Microbiology, Federal University of Rio de Janeiro, Ilha do Fundão, CCS, 21944 - Rio de Janeiro - RJ Brazil

Summary: Five different soils and 3 soil components were treated with bacteriocin produced by A. lipoferum 215. The activity disappeared soon after application in four soils and in humic acid preparation, but persisted in one autoclaved soil poor in organic matter and in clay minerals. A. lipoferum survived for a long period when inoculated into the soil, but no bacteriocin activity was detectable in colonized soils. No bacteriocin activity was found on roots inoculated with the producer and incubated in soil. However, some activity was detected on the inoculated and incubated root after the treatment with Tris-HCl.

Introduction

The term bacteriocin used here describes an inhibitory agent of protein nature active against closely related bacteria. Notwithstanding the widespread occurrence of bacteriocin-producing bacteria and the existence of an abundant literature concerning the isoinhibitory factor, there are very few reports on the ecological role of bacteriocins. The early publications on the production of bacteriocin in natural environments and on their possible role in the regulation of bacterial population dynamics in some ecosystems were discussed in some review articles (1,2,3). Most of the papers reviewed dealt with the bacteriocins produced by pathogenic bacteria or animal-dependent saprophytes. In spite of the fact that the results were not always consistent, there are, however, some indication that bacterial interference caused by bacteriocin activity might play some role in the control of bacterial population in gut or oral cavity. More recently, Hillman et al.(4) showed a durable colonization of the human oral cavity by an introduced bacteriocinogenic strain of Streptococcus mutans, and the effects of the in situ produced bacteriocin on the indigenous S. mutans population.

To the best of our knowledge, production of bacteriocin by nitrogen-fixing bacteria had been examined previously only in rhizobia, and since Roslycky's

Azospirillum IV: Genetics, Physiology, Ecology
Edited by W. Klingmüller

(5) first short report several papers on bacteriocins in Rhizobium have been published. Some of the workers tried to define the ecological role of Rhizobium bacteriocins (among others 6,7,8,9). Schwinghamer & Brockwell (6) found that a bacteriocin producing strain of R.trifolii strongly suppressed the growth of a sensitive strain in broth culture, and that in a sterile peat culture the degree of suppression also was considerable. Hodgson et al. (8) were able to control the nodulation of Trifolium subterraneum by application of a bacteriocin-producing strain of R. trifolii in plant-tube and soil experiments. Information is also available on the effect of bacteriocin-like substances produced by R. trifolii on the nodulation of cowpeas in sterile soil (9). As we have reported, Azospirillum also is able to produce bacteriocin, and the bacteriocin-producing strains are quite common in both A. lipoferum and A. brasilense (10,11,12,13).

The purpose of the assays related here was to check whether the azospirilla are able to produce the bacteriocin in soil and to test the persistence of bacteriocin under soil conditions.

Materials and Methods

Bacteria. Azospirillum lipoferum 215 and A. lipoferum Sp 59 were used throughout this work as producer and indicator, respectively. The bacteria come from the collection of Dr. J. Döbereiner, and the stock cultures were maintained on potato agar (14).

Soils. The characteristics of the soils used are listed on Table 1.

Bacteriocin. Bacteriocin was produced by A. lipoferum 215 grown in nutrient broth, and extracted from the whole cells using a mixture of Tris-HCl 0.05 M, pH 8.1 with 5% NaCl as described by Oliveira (10) and Oliveira and Drozdowicz (12). Solid ammonium sulphate was added to the extract to obtain the final concentration of 55%. The precipitate obtained showed bacteriocin activity and was considered as crude bacteriocin (CB).

Persistence of bacteriocin in soil. Soil aggregates or portions of separated soil components listed on Table 1 were saturated with CB-preparation, containing 0.9 mg protein.ml^{-1}. The treated substrates were kept for 12 days at room temperature. At daily intervals subsamples were taken out, placed on the surface of nutrient agar plates inoculated with indicator strain, incubated at 33ºC, and then examine as to the presence of zones of bacteriocin-like inhibition around the soil spots{modified "spot-test" (11)}. Autoclaved substrates were used with the exception of Red Yellow Podzolic (RYP) soil which was examined both in the natural state and autoclaved. The inhibitory activity of the CB applied was checked before application and non-treated substrate samples were used as controls.

Table 1. Persistence of bacteriocin in soil and in soil components[1].

	Soil organic matter content (%)	pH	Bacteriocin activity[2) Days[3) of incubation 0	1	2	7	12
Garden soil	11.0	6.7	-	-	-	-	nd
Garden soil + 50% SAND	5.0	6.7	-	-	-	-	nd
Red Yellow Podzolic (RYP)	0.9	4.5	+	+	+	+	+
Red Yellow Podzolic(*)	0.9	4.5	-	-	-	nd	nd
Red Yellow Latosol[4)5)]	2.4	4.9	-	-	-	-	nd
Dark Red Latosol[4)5)]	3.2	4.2	-	-	-	-	nd
Humic acid preparation[5)]			-	-	-	-	nd
Bentonite			+	+	+	+	nd
Kaolite			+	+	+	+	nd

1) All the substrates, with the exception of (*) were autoclaved before the treatment with bacteriocin (The (*) soil aggregates were heavily colonized by soil microorganisms on the plates); 2) Presence (+) or absence (-) of zone of inhibition around the sample placed on the surface of nutrient agar plates seeded with indicator A. lipoferum Sp59; 3) 0-Day: tested soon after the application of crude bacteriocin; 4) Soils from the cerrado area; 5) Examined at natural pH and at pH adjusted to 7.0; nd = not determined.

Bacteriocin production in soil. The soils listed in Table 1 were distributed in flasks (appr. 15 g soil per flask). Part of the flasks were autoclaved and another part left in the natural state. A heavy suspension ($2x10^9$ cells.ml^{-1}) of A.lipoferum 215 cells washed and suspended in 0.85% saline was added to the soils in an amount sufficient to obtain a saturation corresponding to 50% of the soil water-holding capacity. The flasks were incubated at room temperature up to three months; the humidity of the soil was controlled and adjusted when necessary. At intervals samples of soil were taken out and examined as to the survival of Azospirillum and the presence of bacteriocin activity. Survival was determined by the Most Probable Number technique, using nitrogen-free,semisolid Nfb medium (14). The bacteriocin activity was checked by the "spot-test". In the experiment with glucose-amended RYP-soil, a 1% glucose water solution was substituted for the saline. In one experiment, 150 g of autoclaved RYP-soil was inoculated and incubated in the above-described manner. After 48 h the soil was mixed with 200 ml of a buffer solution consisting of Tris-HCl 0.05M, pH 8.1 and 5% NaCl. The mixture was agitated for 15 h at 4ºC, twice centrifuged to separate the soil, and the supernatant then precipitated with ammonium sulphate. The precipitate obtained was assayed as to bacteriocin activity by the "spot-test".

Axenic wheat roots were used to test the possibility of bacteriocin production in the root region. Nine to 19 day old axenic roots, provided by the Laboratory of Genetics, were aseptically cut in pieces approximately 1 cm long, heavily inoculated with a suspension of washed producer cells and then partially buried in autoclaved RYP-soil distributed in Petri dishes. Fifteen root pieces were buried in each plate. After 48 h incubation the roots were taken out, the adhering soil removed and then examined as to the presence of bacterial colonization. Part of the colonized roots were treated with chloroform, another part left untreated and then the roots were checked as to bacteriocin activity by a technique similar to that of "spot-test".

All the trials were performed at least in triplicate.

Results and Discussion

The activity of crude bacteriocin introduced into autoclaved soil disappeared soon after application in almost all soils examined (Table 1). The only exception was the autoclaved RYP in which the inhibitory activity was detectable for at least 12 days. In the non-autoclaved RYP the activity disappeared soon after the treatment with CB. The instantaneous disappearance of the introduced activity in most of the autoclaved soils suggests that the inactivation may be attributable to the abiotic soil components, either organic or inorganic. The disappearance of the activity in non-autoclaved RYP confirms the expected participation of microorganisms in bacteriocin degradation. The introduced bacteriocin activity persisted for a relatively long time in clay minerals and in the autoclaved RYP poor in organic matter but disappeared rapidly in humic acid preparation, in latosols and in garden soils. These observations seem to indicate that the soil organic matter is

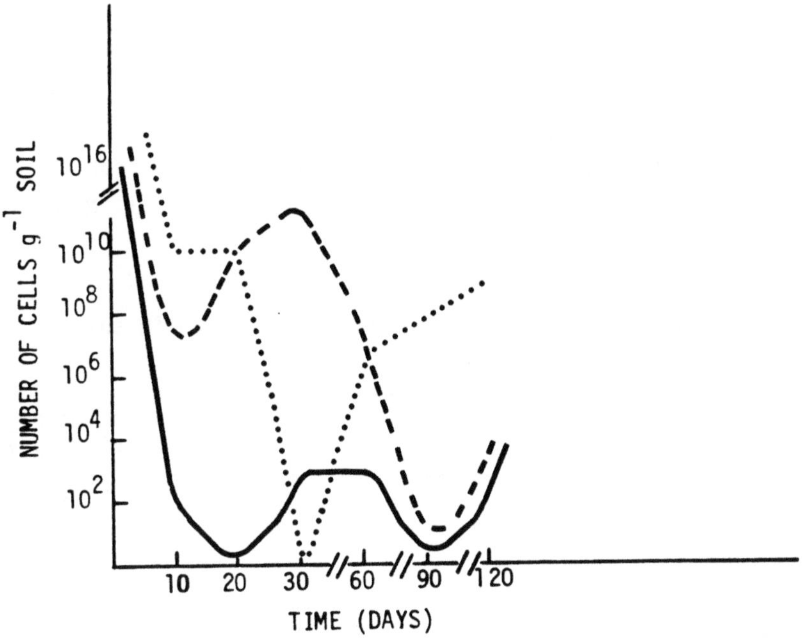

Fig. 1. Survival of Azospirillum in garden soil:(—) native population, (---) natural soil inoculated and (······) autoclaved soil inoculated.

the main factor responsible for the abiotic inactivation of bacteriocin.

Garden soil and RYP contained a native Azospirillum population. Figure 1 shows the behavior of Azospirillum in the garden soil. The introduced A. lipoferum 215 survived in the autoclaved soil for a long time, and in non-autoclaved soil the inoculation increased the general number of azospirilla. In RYP the population dynamics of azospirilla was similar to that exemplified in Fig. 1. The survival of A. lipoferum 215 was also examined in glucose-amended RYP. The general patterns of population dynamics were similar to those observed in unamended soil. In latosols the situation was different: the local bacterial community did not contain any native Azospirillum population, and the introduced A. lipoferum 215 also disappeared soon after inoculation in both natural and autoclaved soils.

Notwithstanding the durable and abundant presence of viable A. lipoferum

215 cells in inoculated garden and RYP soils, no bacteriocin activity was detected in them during the course of the experiment, and obviously no activity was found in latosols (for the sake of brevity no Table was prepared to summarize the negative results). In one experiment, the soil aggregates used in the "spot-test" were replaced by a preparation extracted from the inoculated RYP soil. The protein preparation obtained did not reveal any bacteriocin activity.

Table 2. Behavior of the producer[1] on wheat roots buried in soil.

	Growth[2]	Bacteriocin activity[3]
Colonized root[4]	+	+
Colonized root + $CHCl_3$[5]	-	-
Extract from colonized root[6]	-	-
Colonized root after extraction	-	+
Rhizosphere soil[7]	+	+
Rhizosphere soil + $CHCl_3$	-	-
Controls:		
Tris-HCl 0.05M, pH 8.1 + 5% NaCl	-	-
Axenic root	-	-
Axenic root + $CHCl_3$	-	-
Colonized string	+	+
Colonized strinq + $CHCl_3$	-	-

1) A.lipoferum 215; 2) Detectable colonization of the root by A.lipoferum 215 after 48 h incubation in autoclaved RYP; 3) Presence (+) or absence (-)of the zone of inhibition around the sample placed on the surface of nutrient agar plate seeded with indicator A.lipoferum Sp59; 4) Inoculated root after 48 h incubation in autoclaved RYP; 5) Inoculated root treated with chloroform after removal of the adhering soil; 6) Inoculated root was extracted with Tris-HCl, 0.05M, pH 8.1 + 5% NaCl; 7) Soil removed from the surface of inoculated root.

Table 2 summarizes the results obtained with inoculated wheat roots incubated in autoclaved soil. The roots were heavily colonized by A. lipoferum 215 during the incubation. Distinct zones of inhibition appeared around the colonized roots placed on the nutrient agar seeded with indicator A.lipoferum Sp59. A similar effect was observed when particles of soil liberated from colonized roots (rhizosphere soil) were tested against the indicator. The inhibition did not appear, however, when the colonized roots or the soil were sterilized with chloroform before the "spot-test". This indicates that the inhibition has to be attributed rather to the actual activity of living producer cells introduced with the material tested and grown on the nutrient agar than to bacteriocin produced and accumulated during the incubation in soil. The result obtained with the inoculated, incubated and colonized string, used as one of the controls, corroborates this conclusion.

The Tris extract obtained from the colonized roots did not show any bacteriocin activity. The extracted root, however, inhibited the growth of indicator bacteria (Table 2). This finding suggests that, in spite of all the negative results described above, one can not exclude the possibility that the bacteriocinogenic A. lipoferum 215 might be able to produce bacteriocin in soil on the root surface and that this iso-inhibitor is strongly fixed on the root.

The reisolated producer did not lose its bacteriocinogenity during the proliferation in autoclaved soil and on the root surface.

It is very tempting to assume some ecological role for biologically active compounds such as antibiotics and bacteriocins, and many original papers and review articles on the production and activity of antibiotics in soil have been published. Among sometimes conflicting results there is also well documented information about the role of antibiotics in control of the bacterial community in soil (15,16).

In contrast to the vast literature on the production and activity of antibiotics in soil, there is not, so far as we know, any similar information about bacteriocin. This contribution seems to be one of the first on this topic, and just as during the early period of research on antibiotics in soil, when conflicting results prevailed, our results also are not conclusive.

No bacteriocin production was detected by the technique used in the soil examined. However, three observations were made: the persistence of introduced bacteriocin in autoclaved RYP (Table 1); the presence of bacteriocin activity on the surface of extracted wheat root (Table 2), and finally, the persistence of producer bacteriocinogenity during its proliferation in soil. These three observations suggest that the research on bacteriocin production in soil and on its ecological role deserves to be continued.

Acknowledgements

The authors thank the graduate student Rosangela Maria de Araújo Soares for her participation in this research. The work on Azospirillum has been supported by grants from the Brazilian CNPq and FINEP and from the Commission of the European Community. This support is gratefully acknowledged.

References

1. Reeves,P. (1965). Bacteriol. Rev. 29, 24-45
2. Tagg, J.R., Dajani, A.S. and Wannamaker, L.W. (1976). Bacteriol. Rev. 40, 722-756
3. Koninski, J. (1978). In: Gunsalus, I.C. and Stanier, R.Y. (eds.). The Bacteria, Vol. VI, Ac. Press, New York, San Francisco, London, p. 71-136
4. Hillman, J.D., Yaphe, B.I. and Johnson, K.P. (1985). J. Dent. Res. 64, 1272-1274
5. Roslycky, E.B. (1967). Can. J. Microbiol. 13, 431-432
6. Schwinghamer, E.A. and Brockwell, H. (1978). Soil. Biol. Biochem. 10, 383-387
7. Trinick, M.J. and Parker, C.A. (1982). Soil Biol. Biochem. 14, 79-86
8. Hodgson, A.L.M., Roberts, W.P. and Waid, J.S. (1985). Soil Biol. Biochem. 17, 475-478
9. Joseph, M.V., Desai, J.D. and Desai, A.J. (1985). Appl. Environ. Microbiol. 49, 459-461
10. Oliveira, R.G.B. (1987). PhD. Thesis, Fed. Univ. Rio de Janeiro
11. Oliveira, R.G.B. and Drozdowicz, A. (1981). Rev. Microbiol. (S. Paulo) 12, 42-47
12. Oliveira, R.G.B. and Drozdowicz, A. (1987). Zblt. Mikrobiol. 142, (in press)
13. Skorupska, A., Brzezinska, M., Choma, A., Kulinska, D. and Lorkiewicz, Z. Microbiols, 44, 243-251
14. Döbereiner, J., Marriel, I.E. and Nery, M. (1976). Can. J. Microbiol. 10, 1464-1473
15. Coelho, R.R.R. and Drozdowicz, A. (1978). Rev. Ecol. Biol. Sol. 15, 459-473
16. Döbereiner, J., Scotti, M.R.M.M.L., Sã, N.M. and Vargas, M.A.T. (1981),in Current Perspectives in Nitrogen Fixation, Australian Acad. Sci., p. 434.

NITROGEN FIXING ABILITY OF AZOSPIRILLUM LIPOFERUM IN A SPERMOSPHERE SYSTEM IN RELATION TO FORMER GROWTH CONDITIONS.

Debendra Nath Nayak
Botany Department, J.K.B.K. College,
Cuttack-753 003, Orissa, India.

Summary: Cells of an *Azospirillum lipoferum* strain isolated from rice roots were grown in both nitrogen fixing condition [in Enriched malate broth in an atm. containing 1% O_2, 30% N_2 and 69% Ar] and non fixing condition [in nutrient broth] to check their associative ability with the rice roots. After growing the bacteria in two different conditions the associative nitrogen fixing ability of the inoculum was determined after inoculating them to etiolated sterile seedlings of two rice varieties [IR42 and OS4] in the spermosphere system of Thomas-Bauzon *et al* [1]. The result did not show any difference between the both kinds of inoculum. Keywords: Nitrogen fixation, *Azospirillum lipoferum*, spermosphere, rice, etiolated seedlings.

Introduction

To study the associative nitrogen fixing ability between the plant and bacteria *in situ* the spermosphere model of Thomas-Bauzon *et al*. [1] offers a useful tool. In spite of its draw back the technique is certainly handy to screen the various plant bacterial association. In the present study I therefore wanted to check the dogma as to in which condition the inoculum should be grown i.e. in nitrogen fixing or in non nitrogen fixing conditions, as the survibility and the colonizing ability of the inoculum are two vital points for the success of inoculation experiments.

Azospirillum IV: Genetics, Physiology, Ecology
Edited by W. Klingmüller

Materials and Methods

Seed sterilization and germination

Seeds of OS4 and IR42 were taken. The dehulled seeds of uniform shape, color, opacity and mass were sterilized with a succession of disinfectants and fungicides [200 ml agent for every 100 seeds] as follows: 70% ethyl alcohol for 5 minutes, undiluted commercial hypochlorite bleach preparation for 20 mn., and finally, acidified mercuric chloride [0.1%] for 2-3 minutes. Prior to transfer into each succeeding agent, the seeds were washed of the preceeding disinfectants two to three times with sterile distilled water, and after the last agent, 20 times with the same. To detect the associative nitrogen fixation the spermosphere technique of Thomas-Bauzon *et al.* [1] were tried. The surface sterilized seeds were transferred to silicon-plugged sterile culture tubes [pyrex, 150 x 16 mm O.D.] containing 5 ml of a C - and N-free semisolid medium used by Thomas-Bauzon *et al.* [1]. The seeds were allowed to germinate in the dark at 30°C until the coleoptiles and radicles were 1-2 cm long [about 3-5 days depending on the variety].

Preparation of the inoculum

Bacterial strain of *Azospirillum lipoferum* was cultivated in Erlenmeyer flasks containing 40 ml of enriched nutrient broth [ENB] for 18-20 h with constant shaking at ambient temperature. The cells were harvested by centrifugation at 8000 rpm for 15 minutes and washed thrice in sterile phosphate buffered saline. The washed cells were reinoculated to fresh ENB for growing under non fixing condition and to enriched malate broth [EMB] for growing under nitrogen fixing condition.

Growth under non fixing conditions

The cells were grown as usual in ENB and harvested in 0.01 M phosphate buffered saline [pH 7.0]. The optical density [450 nm] of the resulting cell suspension was adjusted to a predetermined concentration of 6×10^8 cells/ml. The viable count was made on congored medium.

Growth under nitrogen fixing conditions

For growing the cells under nitrogen fixing condition 125 ml quantities of the malate broth [2] enriched with yeast extract [EMB] but without agar was taken in 250 ml Erlenmeyer flasks. After sterilization at 15 lbs pressure for 15 minutes the medium was inoculated with 1 ml quantities or washed bacterial suspension previously grown in ENB. The cotton plugs were immediately changed by UV sterilized rubber corks with a special part of introducing gas mixture. Each flask was evacuated completely by evacuating and flushing with argon alternatively for 3 times. After 3rd evacuation gas mixture containing 1% 0_2, 30% N_2 and 69% Ar. was introduced into each flask through sterile cotton filters. All the flasks were incubated at 32°C for 20 hours.

The cells were harvested similarly as for ENB and the O.D. was adjusted to 9 x 10^8 cells/ml at 450 nm. The viable count were also made in the similar way as mentioned before.

Measurement of ARA

Seedings of OS4 and IR42 showing uniform development and no contamination were selected and the rest discarded. Seedlings were inoculated with 0.2 ml of cell suspension to give 10^7 cells. Immediately after inoculation the tubes were transferred into larger culture tubes[200 x 25 mm O.D] containing 3 ml of 0.1 N NaOH and sealed with Suba-seal No.49 sleeve-type rubber stoppers. Three milliliter of the air inside the outer tube was replaced with 3 ml of a gas mixture consisting of pure acetylene [99% CIGI] and propane in a 500:10 proportion to give approximately 5% C_2H_4 atmosphere in the tube. Control tubes without seedling and without bacteria were also taken. The tubes were incubated in the dark at 30°C. Ethlene and propane was measured at 0,3,6 and 10 days using a Hitachi FID gas chromatograph equipped with a Porapak N column [1m x 4mm O.D].

Results and Discussion

Nitrogen fixation was found to occur in close association of rice plant; planted hills were shown to have considerably greater activity than unplanted hills by *in situ* ARA measurement [3,4]. From a laboratory study the same trend was observed when different groups of bacteria isolated from rice roots gave no activity without the rice seedling. The results are shown in Table 1. In spermosphere model both the partner get an opportunity to interact. The germinating rice releases organic compounds which the bacteria utilize for growth and energy source for its nitrogen fixing machinery. Uptake of the fixed nitrogen by the growing plant and growth of bacteria maintains the nitrogen-deficient conditions which enables such activity to be expressed.

For preparation of inoculum, it is very essential to find out the suitable way of cultivating the inoculum. Whether the inoculum acquires more competitive ability and better colonizing capacity when grown in nitrogen fixing conditions or non nitrogen fixing condition. It is well known when *Azospirillum* sp. are grown in nitrogen fixing conditions show the

Table 1. Nitrogen fixation associated with OS4 seedlings grown by the spermosphere technique and inoculated with various genera of bacteria.

Treatment	n moles C_2H_4/Plant/7 day*	
	+ Seedling	- Seedling
Control	0.10	0.17
A.Lipoferum 34H	1500 ± 336	< 0.17
Pseudononas H8	900 ± 267	< 0.17
A brasilense 26H2	318 ± 213	< 0.17
K Cloacae EF 26-1	78 ± 6	< 0.17

* Mean ± SEM of 5 replications.

presence of large amount of reserve fat granules [PHB granules] within the cells. These reserve fat bodies might give better colonizing capacity or aggressiveness to the organisms to establish themselves in adverse conditions.

To test the fact, the spermosphere model was tried. The data presented in Table 2 showed that the Azospirillum grown in both type of situation had equal associative nitrogen fixing ability. However, the nitrogen fixing capacity of IR42 was slightly higher than the OS4 in the spermosphere. The result however should be repeated in various other varieties and bacterial strains. Another important factor which has a major role the plant-bacterial association is soil. The spermosphere systems surely neglects the infuence exerted by the soil on the interaction between bacteria and plant.

The infected root of both the varieties inoculated with bacteria grown in two different conditions were stained with phenol aniline blue and were examined under a light microscope. It showed that Azospirillum strain occured in clumps and were not evenly distributed, especially on the root cap and base of lateral root branches but there was absolutely no difference in the colonising behaviour of EMB and ENB grown strains.

Table 2. The associative nitrogen fixing ability of two rice varieties inoculated with Azospirillum lipoferum grown in nitrogen fixing and non fixing conditions using spermosphere model.

Variety	Treatment	n moles C_2H_4/tube*		
		3rd day	6th day	10th day
OS4	ENB	2956 ± 423	3990 ± 305	4661 ± 258
	EMB	3177 ± 463	4218 ± 413	4760 ± 135
IR42	ENB	4458 ± 270	5396 ± 265	5795 ± 105
	EMB	4354 ± 367	4760 ± 162	6040 ± 341

* Mean ± standard deviation of 10 replications

Acknowledgement: I wish to thank the International Rice Research Institute Philippines to provide me with the necessary facilities.

References

1. Thomas-Bauzon, D., Weinhard, P., Villecourt, P. and Balandreau, J. [1982]. Can. J. Microbiol, 28, 922-928

2. Neyra, C.A. and Van Berkam, P. [1977]. Can. J. Microbiol, 23, 306-310

3. Yoshida, T. and Ancajas, R.R. [1973]. Soil Biol. Biochem. 5, 153-155

4. Watanabe, I., Lee, K.K., Alimangro, B.V., Sato, M., del Rosario, D.C. and de Guzman, M.R. [1977]. IRRI paper series No.3, Los-Banos, Phil, 16p.

N_2-DEPENDENT GROWTH OF *AZOSPIRILLUM* SPP. IN BATCH CULTURES AT LOW CONCENTRATIONS OF OXYGEN

T. Hurek, B. Reinhold, E.-G. Niemann, and I. Fendrik
Institut für Biophysik, Universität Hannover, Herrenhäuser Str. 2, 3000 Hannover 21, FRG

Summary: *Azospirillum halopraeferens* Au7 and *A. amazonense* Y1 were grown in batch cultures under controlled conditions on N_2 as substrate at about 1 µM dissolved O_2 concentration, both strains showed exponential growth up to an optical density of 13 for Y1 on 10 g/l of fructose (2.9 g bacterial dry weight per liter) and of 3 for Au7 on 5 g/l of L-malate (900 mg bacterial dry weight per liter). The generation time of Au7 was 4.2 h, and for Y1 13.5 h. Although the cells grew strictly exponentially, physiological changes could be seen during this growth phase. Protein contents of cells increased from 43 to 70% for Au7 and from 22 to 33% for Y1, efficiency of nitrogen fixation increased from 4 to 35 mg N per g L-malate for Au7 and from 14 to 27 mg N per g fructose for Y1. At the time of highest protein contents 10% of fixed nitrogen was found in the culture filtrate. Since no soluble protein was observed in the supernatant, the combined nitrogen was likely to be due to nitrogen export and not to cell lysis. This may give some evidence for the potential of *Azospirillum* to excrete nitrogen. Keywords: *Azospirillum*, efficiency, N-export

Introduction

The studies on the growth physiology of *Azospirillum* so far reported dealt with *A. lipoferum* and *A. brasilense*, there is only little information available on corresponding properties of *A. amazonense*, and nearly no information in the case of *A. halopraeferens*.

It has been shown by Volpon et al. (1) that in batch cultures of *A. lipoferum* BR17 25% of fixed nitrogen in forms of soluble organic compounds was recovered in the supernatant. For other free-living diazotrophs and also the *Sesbania Rhizobium* ORS571 usually 10% of fixed nitrogen is found in the supernatant of N_2 grown cells (2, 3). Other *Rhizobium* strains export fixed nitrogen into the supernatant in much higher amounts like the cowpea strain

Azospirillum IV: Genetics, Physiology, Ecology
Edited by W. Klingmüller

CB765 which has been reported to leave about 90% of fixed N in the medium (4).

The increase of nitrogen in culture supernatant at the end of logarithmic growth was accompanied by an increase of the utilization of the carbon substrate for nitrogen fixation up to 48 mg N_2 fixed per g glucose consumed (1). Differences in efficiency are also known for other free-living diazotrophs, the efficiency being highest at low oxygen concentrations and at the end of logarithmic growth. For *Azospirillum brasilense* Sp7, efficiencies up to 92 on lactate have been reported (5). In this study we report some features of N_2-fixing *A. amazonense* and *A. halopraeferens* in batch cultures under well controlled conditions at optimal O_2 concentrations with special emphasis on nitrogen distribution and nitrogen fixation efficiency.

Materials and Methods

Precultures of *Azospirillum halopraeferens* Au 7 (ATCC 43710) were grown on combined nitrogen in sythetic malate (SM) medium (6) modified by the addition of 2.5 g/l of NaCl and adjustment of the pH to 7.2. For *A. amazonense* Y1 (LMG 7115), the SM medium was modified by lowering the pH to 6.0, omitting the yeast extract and replacing the sodium malate by 10 g/l of fructose. Both strains were incubated at 37°C.

N_2-dependent growth of dissolved-O_2-controlled batch cultures was carried out in a 2 l fermentor (2) by using the same media mentioned above without combined nitrogen. Maintenance of the desired partial pressures of oxygen was done as described previously (2). The culture vessel was inoculated with 5ml of a mid-exponential-phase preculture, which had been washed twice with the suitable N-free medium.

Optical density of the cell suspensions was measured at 578 nm using a Beckmann spectrophotometer 25. Bacterial dry weight was determined after freeze drying (7). Cellular protein was determined by the micro-Goa method (8), soluble protein in the culture filtrate by the Bio-Rad protein assay by using bovine serum albumin as a reference standard. L-Malate was assayed enzymatically by a test combination purchased from Boehringer Mannheim GmbH. Fructose was also determined enzymatically (9). Total nitrogen and NH_4^+ were determined as described previously (10).

Results and Discussion

A logarithmic growth response of both *Azospirillum* strains was obtained at optimal O_2 concentrations, as shown by three different parameters each (Fig.1, Fig. 5). Growth stopped when the fructose or malate concentration in the medium was below 1 mg per liter (Fig. 2; data for Au7 not shown)

The growth rate for Y1 was 0.05 h^{-1} and the generation time 13.5 h, which corroborated the observation that *A. amazonense* grows much slower than the other *Azospirillum* species (11). For Au7 a growth rate of 0.16 h^{-1} and a generation time of 4.2 h was calculated from the growth curve. Among root-associated microaerophilic N_2-fixing microorganisms this strain and *A. lipoferum* Rp5 (growth rate on L-malate $\geq$ 0.17 h^{-1}; 2) also isolated from the root-surface of Kallar grass roots (10, 12) show the highest growth rates so far reported.

Molar growth yields of 24 and 52 g dry weight per mol L-malate and fructose were estimated for Au7 and Y1 respectively which corresponded well with data from Kloss et al. (7) for *A. brasilense* Sp7 (26.8 g dry weight per mol malate) and with data from Volpon et al. (1) for *A. lipoferum* BR17 (46 g dry weight per mol glucose).

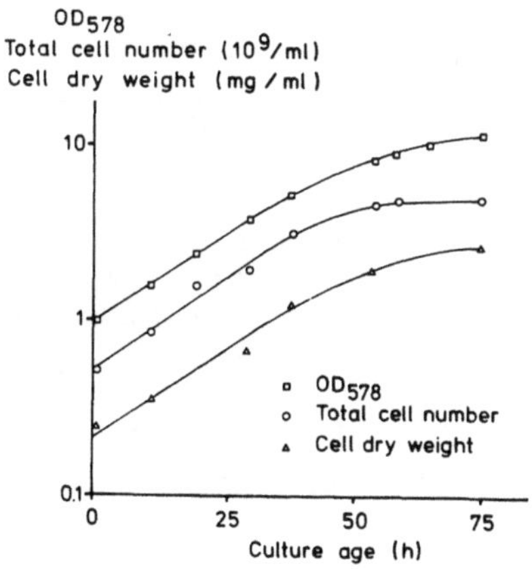

Figure 1 N_2-dependent growth of *A. amazonense* Y1 with 10 g/l of fructose. Parameters controlled: stirring rate 400 rpm, 37°C, pH 6.0, dissolved oxygen 2µM.

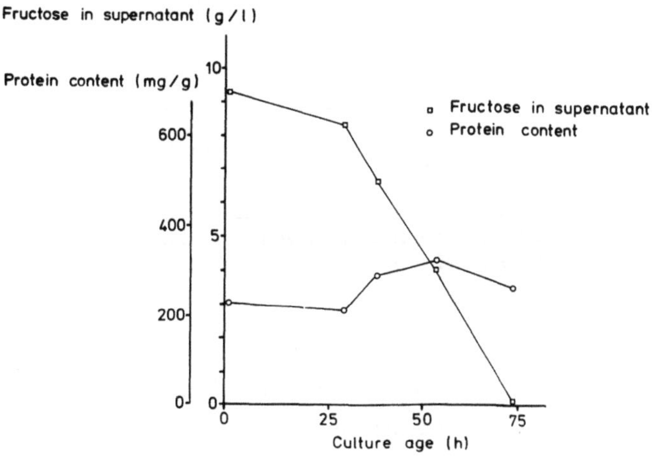

Figure 2 Remaining fructose in supernatant and protein content of cells in batch culture of *A. amazonense* Y1. Culture conditions were as described in Fig. 1.

Although the cells grew strictly exponentially, physiological changes could be seen during logarithmic growth, which were in good agreement with corresponding data from Volpon et al. (1). Protein contents of cells increased from 22 to 33 % for Y1 (Fig. 2) and similarly from 43 to 70 % for Au7. Efficiency of nitrogen fixation increased from 4 to 35 mg N per g L-malate for Au7 (Fig. 5) and from 14 to 27 mg N per g fructose for Y1 (Fig. 3) at a time when the C-source was clearly not limiting. At the time of highest efficiency still about 8 g fructose and 3 g L-malate per liter were in the media of the respective strains. Thus relatively high efficiencies at low O_2-concentrations can be reached when growth is not carbon limited in agreement with results obtained for *Xanthobacter autotrophicus* (13).

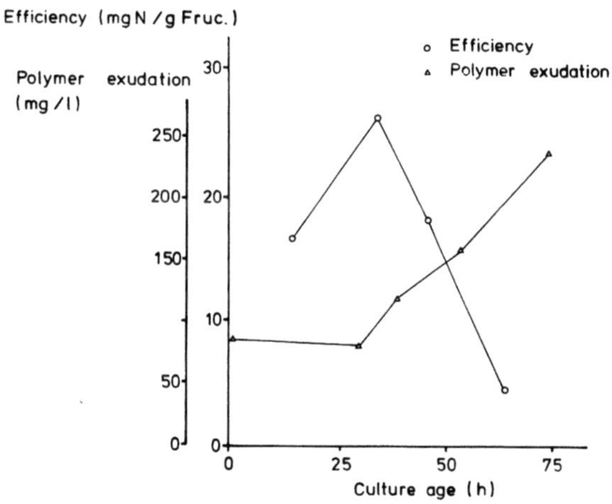

Figure 3 Efficiency of nitrogen fixation and polymer exudation in batch cultures of *A. amazonense* Y1. Culture conditions as described in Fig. 1.

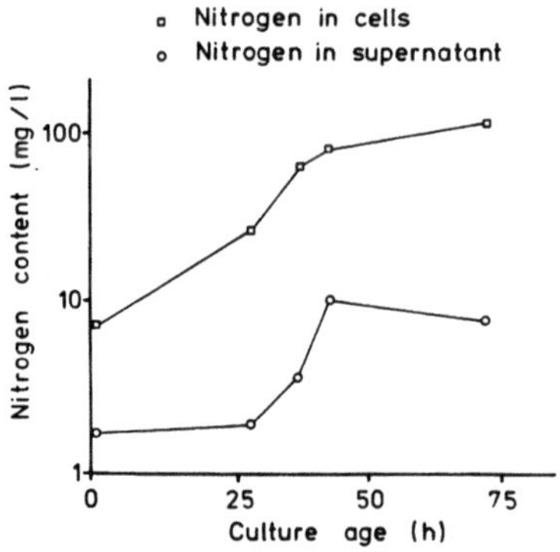

Figure 4 Distribution of fixed nitrogen in batch cultures of *A. amazonense* Y1. Culture conditions as described in Fig. 1.

Furthermore, in cultures of Y1 the occurence of a polymer was observed, which accumulated at the end of logarithmic growth in the supernatant. The nature of this substance is not known.

At the time of highest protein contents, loss of fixed nitrogen into the culture filtrate of both strains was also highest and resulted in about 10% of fixed nitrogen in the culture filtrate (Fig. 4; similarly for Au7, 7 mg N from 69 mg N fixed were found in the supernatant). Since the concentration of soluble protein in the culture filtrates were negligible (≤1 µg/ml) the combinec nitrogen was likely to be due to nitrogen export and not to cell lysis.

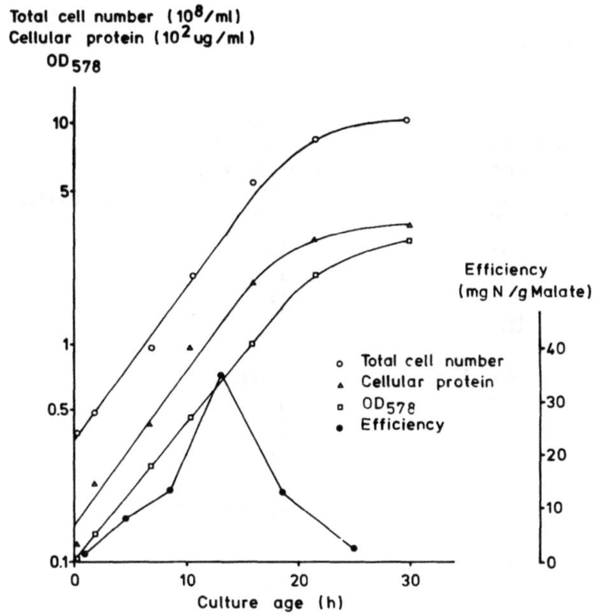

Figure 5 N_2-dependent growth of *A. halopraeferens* Au 7 with 5 g/l L-malate. The following parameters were contolled: stirring rate 400 rpm, temperature 41°C, pH 7.2, dissolved oxygen 1 µM O_2.

In N_2-grown chemostat cultures protein contents and efficiencies are usually highest at low O_2-concentrations, and in NH_4^+-grown chemostat cultures under air saturation protein contents are usually lower (Table 1). Thus the increase of protein content and of efficiency observed at low O_2 concentrations in batch cultures of *Azospirillum* spp. may reflect a shift of the cells: the cells originating from a preculture grown under air saturation on NH_4^+ may adapt their physiological state during logarithmic growth to the low optimal O_2 concentration applied.

However, in N_2-grown, C-limited steady-state cultures of free living diazotrophs combined nitrogen in supernatant can usually not be attributed to excretion, furthermore efficiencies found are not as high as observed in batch cultures under comparable low O_2 concentrations. The reasons for the physiological state resulting in nitrogen excretion are not yet understood.

TABLE 1. Effect of oxygen on percent protein of bacterial dry weight and efficiency of N_2-fixation of various diazotrophs grown on N_2 at a dilution rate of 0.08 h^{-1} at various dissolved O_2 concentrations.

Organisms	dissolved oxygen concentration (μM)	percent protein of bacterial dry weight (%)	efficiency of nitrogen fixation mg N fixed / g L-malate consumed
	1.7	45	12.5
A. brasilense[a]	7.6	27	8.0
Sp7	13.2	14	0.8
	16.0	8	0.9
	0.6(190)[c]	66(49)[c]	16.5
A. lipoferum[b]	2.2	43	11.0
RP5	9.0	31	6.7
	11.6	25	1.0
	0.8(170)[c]	80(39)[c]	20.0
A. halopraeferens[b]	1.3	74	17.5
Au4	2.7	45	12.0
	4.8	34	10.0
	0.3	73	15.0
	1.1(163)[c]	80(74)[c]	14.5
diazotrophic rod[b]	3.8	67	12.5
BH72	7.7	62	12.5
	14.9	38	8.9
	27.2	33	6.2
	0.3	56	9.4
diazotrophic rod[b]	1.4(169)[c]	58(65)[c]	6.3
H6a2	10.8	61	4.5
	22.1	63	4.3
	30.8	43	4.0

[a] Unpublished results. Methods used as described in reference (2).
[b] Data from reference (2).
[c] Data in parentheses included for comparison. Cultures grown on 1 g NH_4Cl / l. Hurek, unpublished results. Methods used as described in reference (2).

References

1. Volpon, A. G. T., De-Polli, H., and Döbereiner, J.(1981). Arch Microbiol. 128, 371-375
2. Hurek, T., Reinhold, B., Fendrik, I., and Niemann, E.-G.(1987). Appl. Environ. Microbiol. 53, 163-169

3. Gebhardt, C., Turner, G. L., Gibson, A. H., Dreyfus, B. L., and Bergersen, F. J.(1984). J. Gen. Microbol. 130, 843-848
4. Bergersen, F. J., and Turner, G. L.(1978). Biochim. Biophys. Acta 538, 406-416
5. Stephan, M. P., Pedrosa, F. O., and Döbereiner, J.(1981). In: Vose,P. B., and Ruschel, A. P.(eds.), Associative N_2-fixation, CRC Press, Boca Raton, Florida, vol. 1, 7-13
6. Reinhold, B., Hurek, T., and Fendrik, I.(1985). J. Bacteriol. 162, 190-195
7. Kloss, M., Iwannek, K.-H., and Fendrik, I.(1983). J. Gen. Appl. Microbiol. 29, 447-457
8. Bergersen, F. J.(1980). In: Bergersen, F. J. (ed.), Methods for evaluating biological nitrogen fixation, John Wiley and Sons, Chichester, 65-110
9. Bernt, E., and Bergmeyer, H. U.(1970).In: Bergmeyer, H., U. (ed.), Methoden der enzymatischen Analyse, Verlag Chemie, Weinheim, vol. 2, 1266-1269
10. Reinhold, B., Hurek, T., Niemann, E.-G., and Fendrik, I.(1986). Appl. Environ. Microbiol. 52, 520-526
11. Döbereiner, J.(1983). In: Klingmüller, W. (ed.), *Azospirillum* II, Genetics, Physiology, Ecology, Birkhäuser Verlag, Basel, Experientia Supplementum 48, 9-23
12. Reinhold, B., Hurek, T., Fendrik, I., Pot, B., Gillis, M., Kersters, K., Thielemans, S., and De Ley, J.(1987). Int. J. Syst. Bacteriol. 37, 43-51
13. Berndt, H., Ostwal, K.-P., Lalucat, J., Schumann, C., Mayer, F. and Schlegel, H. (1976). Arch. Microbiol. 108, 17-26

OSMOREGULATORY PROPERTIES OF AZOSPIRILLUM SPP.

Anton Hartmann
Lehrstuhl für Mikrobiologie, Universität Bayreuth
Universitätsstraße 30, D-8580 Bayreuth, FRG

Summary: Within the genus *Azospirillum*, the osmotolerance increased in the order *A. amazonense*, *A. lipoferum*, *A. brasilense*, *A. halopraeferens*. *A. halopraeferens* specifically required sodium ions (10 mM) for growth. *A. brasilense* and *A. halopraeferens* utilized amino acids (such as glutamate and proline) poorly, glycine betaine and choline not at all as nitrogen and carbon sources. At osmotic stress conditions, growth and nitrogen fixation of *A. brasilense* and *A. halopraeferens* were stimulated by low concentrations of glycine betaine and choline. A high affinity uptake scavenged these osmoprotective substances from the environment and caused their intracellular accumulation. Spontaneous mutants of *A. brasilense* Sp 7, resistant to the proline analogue 3,4-dehydroproline, were more osmotolerant as compared to the wild type. Keywords: Osmoregulation, amino acids, betaines, *A. amazonense*, *A. lipoferum*, *A. brasilense*, *A. halopraeferens*.

Introduction

In the osmoregulation of microorganisms amino acids and betaines play an important role as compatible solutes. Their intracellular accumulation compensates low water potentials in the environment at water stress conditions (e.g. drought, salt stress), prevents water loss from the cytoplasm and preserves enzymatic functions (1, 2, 3). In drought or osmotic stress conditions various plants, including grasses, accumulate betaines and/or proline in their tissues (4). These substances are probably available for the microorganisms in the rhizosphere.

In its natural environment, *Azosprillum* has to cope with water

Azospirillum IV: Genetics, Physiology, Ecology
Edited by W. Klingmüller

stress in salt affected soils or when soil gets dry. In the rhizosphere, water stress can occur due to the high water needs of the transpiring plant, especially in dry climates. The possible application of Azospirillum inoculations for crop improvement under semiarid conditions (5) and in salt affected soils call for a better knowledge of the osmoregulatory properties of Azospirillum spp. and how its environmental fitness can be improved.

Recently, a salttolerant Azospirillum species, A. halopraeferens, was described (6). Among other N_2-fixing bacteria, members of this new species were isolated from the rhizoplane of the halophyte kallargrass (Leptochloa fusca) in Pakistan, which produces 20 - 40 tons of green fodder per ha per year without nitrogen fertilizer added (7, 8). A. halopraeferens was included in the present comparative study of osmoregulatory properties of Azospirillum.

Material and Methods

The Azospirillum strains were obtained from the Deutsche Sammlung für Mikroorganismen (Göttingen): A. brasilense Sp7 (DSM 1690^T), SpCd (DSM 1843), JM6A2 (DSM 1858), SpP2 (DSM 2296), SpT60 (DSM 2298); A. lipoferum Sp59b (DSM 1691^T), SpRG20a (DSM 1840), SpBr17 (DSM 1842), SpA3a (DSM 1838), SpCol3 (DSM 2294) and A 1-3 (DSM 1726); A. amazonense Y1 (DSM 2787^T), Y2 (DSM 2788) and Y6 (DSM 2789). A. halopraeferens Au 4 and Au 5 were kindly provided by B. Reinhold, A. brasilense Sp245 by Dr. J. Döbereiner and A. brasilense Nig5a and OP2 by Dr. F. Favilli.

The Azospirillum strains were grown in liquid mineral medium with ammonium or in semisolid, nitrogen free medium as described (6, 9). MOPS-buffer (100 mM, pH 7.2) was included in the medium of A. halopraeferens (6) to stabilize the pH. A. halopraeferens cultures were incubated at 42°C; the other species were grown at 30°C. Nitrogenase assays were performed with the acetylene reduction technique (10). Growth was recorded by measuring the absorbance at 560 nm. All reagents used were of high purity grade

and obtained from Merck or Sigma.
Abbreviations: CCCP, carbonyl cyanide m-chlorophenylhydrazone; HQNO, 2-heptyl-4-hydroxyquinoline-N-oxide; MOPS, 3-(N-morpholino) propanesulfonic acid; DHP, 3,4-dehydroproline.

Results and Discussion

In *Azospirillum*, the osmotolerance declined in the order *A. halopraeferens*, *A. brasilense*, *A. lipoferum*, *A. amazonense* (Table 1).

Table 1: Effect of sodium chloride on growth of *Azospirillum* spp. in nutrient broth medium

Strains	Generation time (hours) without NaCl	+ 0.1 M NaCl	+ 0.3 M NaCl	+ 0.5 M NaCl	+ 0.8 M NaCl
A.halopraeferens Au4	5.7	3.6	3.5	4.5	39
A.brasilense Sp7	3.0	3.0	7.8	14.8	80
A.lipoferum SpBr17	2.5	3.0	13.5	31.0	n.t.
A.amazonense Y1	2.0	3.0	12.0	no growth	n.t.
" " Y2	2.3	2.8	10.8	no growth	n.t.
" " Y6	2.3	3.0	8.8	33.0	n.t.

n.t.: not tested

The same order of osmotolerance was also found in minimal medium with ammonium as nitrogen source (not shown) and in nitrogen free semisolid medium (Fig. 1). This confirms and extends the reports of Tarrand et al. (11), who found a better tolerance towards NaCl of *A. brasilense* strains as compared to *A. lipoferum* strains, and of Reinhold et al. (6). As candidates to improve osmotolerance, the amino acids glutamate and proline as well as glycine betaine were tested. The addition of glutamate or proline (5 mM) and glycine betaine (1 mM) to minimal medium resulted in better

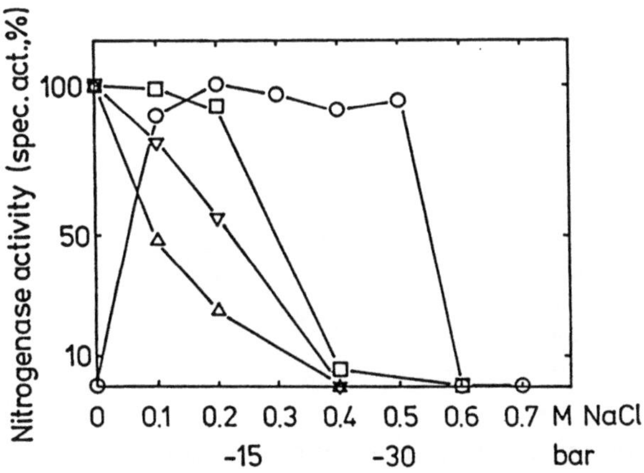

Figure 1: Influence of sodium chloride on the nitrogenase activity of A. amazonense Y1 (△), A. lipoferum SpBr17 (▽), A. brasilense Sp7 (□) and A. halopraeferens Au4 (○). Acetylene reduction was measured in semisolid medium cultures 48 hours after inoculation with 5 % (v/v) stationary grown minimal medium cultures.

growth at inhibitory NaCl concentrations in A. brasilense and A. halopraeferens, but not in A. lipoferum and A. amazonense (not shown). The addition of glycine betaine or choline did not stimulate nitrogen fixation of A. amazonense and A. lipoferum at osmotic stress (Fig. 2). However, betaine glycine (N-trimethylglycine) and N-dimethylglycine stimulated nitrogen fixation in A. brasilense and A. halopraeferens at otherwise totally inhibitory osmotic stress (Fig. 2). Choline could only be used by A. halopraeferens for osmoregulatory purposes. This was confirmed by the finding that only A. halopraeferens had a high affinity uptake for choline (K_M: 16 µM).

To find the physiological basis for the different osmoprotective potential of amino acids and betaines in Azospirillum spp., the utilization of glutamate, proline, glycine betaine and choline as sole nitrogen and carbon sources was tested (Fig. 3). A. amazonense and A. lipoferum used glutamate and proline most efficiently as carbon and nitrogen sources, while A. brasilense

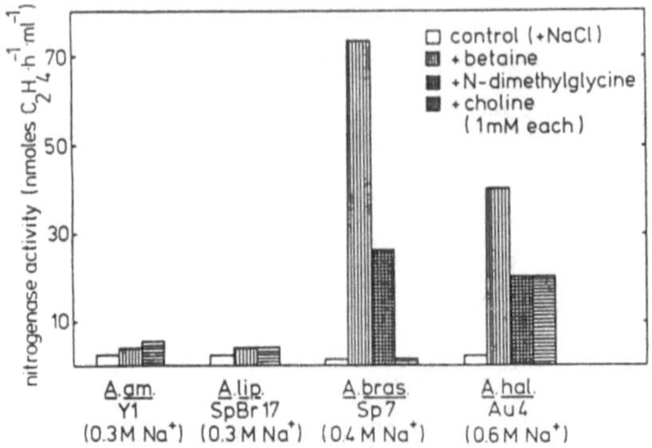

Figure 2: Influence of glycine betaine and related compounds on the specific nitrogenase activity of Azospirillum spp. at osmotic stress conditions.

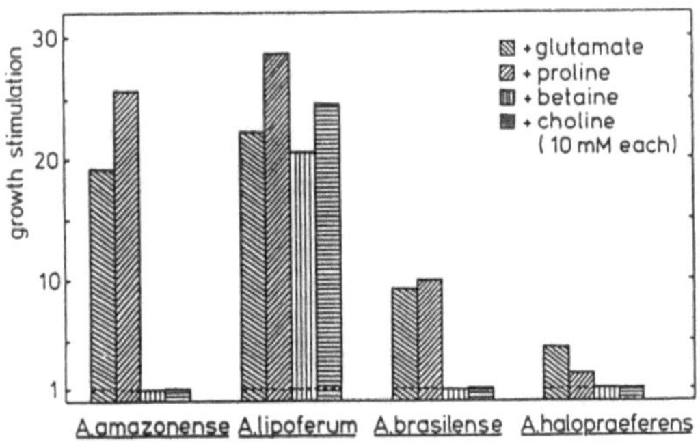

Figure 3: Growth of Azospirillum spp. with amino acids and betaines as sole carbon and nitrogen sources. Growth stimulation was determined by measuring growth ($A_{560\ nm}$) in semisolid cultures after 72 hours incubation with amino acids or betaines, relative to the A_{560} in control cultures without additions.

and even more A. halopraeferens did this less efficiently. A. amazonense (three strains tested) could not grow on betaine and choline, because it cannot take up these substances (not shown). A. lipoferum (six strains tested) readily grew with betaine and choline as nitrogen and carbon sources (Fig. 3). A. brasilense (eight strains tested) and A. halopraeferens (two strains tested) could not grow on glycine betaine and choline. The low utilization of amino acids and betaines for growth correlated well with the potential of these substances as osmoregulators. A. brasilense and A. halopraeferens possibly accumulate betaines occurring in the rhizosphere of osmotically stressed plants (12) to diminish their own water stress.

The salt dependency of A. halopraeferens was investigated in more detail. A. halopraeferens could not grow in minimal medium in the absence of NaCl. The addition of 10 mM NaCl was required for growth and nitrogen fixation (Fig. 4). KCl could not replace

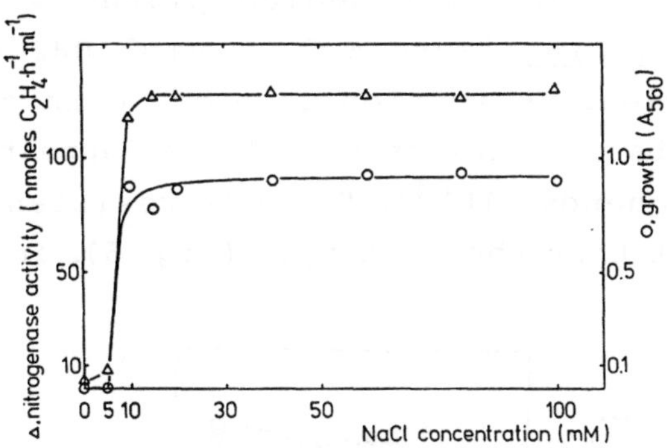

Figure 4: Sodium dependency of growth (O) and nitrogenase activity (Δ) of A. halopraeferens Au4. Semisolid cultures with different additions of NaCl were incubated for 48 hours before the acetylene reduction assay was performed.

NaCl, but Na_2CO_3 and Na_2SO_4 had the same stimulatory effect (not shown). Polyethylenglycol (PEG 400) and glycerol had no stimulatory effect on A. halopraeferens at the same osmotic value as NaCl (100 mM). As an adaptation to the sodium rich alkaline soil, certain cellular processes (e.g. transport processes) of A. halopraeferens may require sodium ions as is the case for the halotolerant soil bacterium Ba 1 (13) or marine bacteria. However, no evidence for a primary sodium pump as in Vibrio alginolyticus

(14) was found (A. Hartmann, M. Ochs, unpublished results). Growth of A. halopraeferens was sensitive towards CCCP (10 µM), which indicated that sodium could not replace protons in coupling respiration and energy production. In addition, the membrane bound NADH oxidase was only slightly stimulated by sodium and potassium ions and was not inhibited by CCCP and HQNO.

The toxic proline analogue 3,4-dehydroproline (DHP) was used to select for mutants with improved osmotolerance. Resistance to DHP could be caused by an increased cellular proline pool, which conferred enhanced osmotolerance in Salmonella typhimurium (15). In contrast to A. halopraeferens, A. brasilense strains were very sensitive to DHP. On minimal medium plates with 0.2 M NaCl, inoculated with A. brasilense Sp7, 2 µmol DHP added into the center caused a clear inhibition zone of 70 mm diameter. After several days incubation, resistant colonies appeared spontaneously with high frequency (10^{-4}). The DHP-resistant mutants were more osmotolerant than the wild type (Fig. 5). In nutrient broth

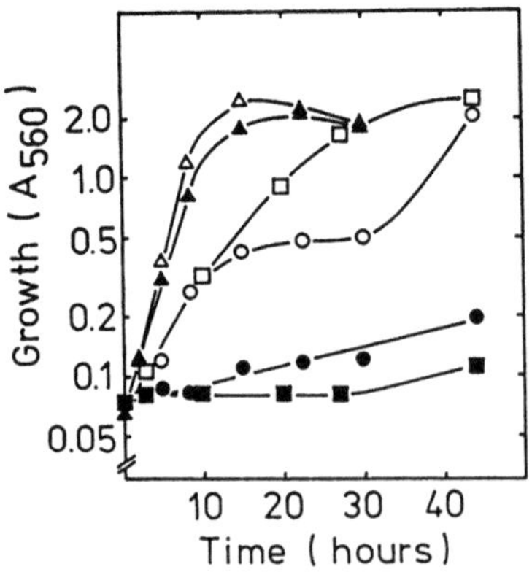

Figure 5: Comparison of the osmotolerance of A. brasilense Sp7 (closed symbols) and a DHP-resistant mutant DHP4 (open symbols). The cultures were grown in minimal medium without NaCl added (▲△), with 0.3 M NaCl added (●○) and with 0.3 M NaCl and 1 mM glycine betaine added (■□).

medium containing 0.5 M NaCl, a DHP-resistant population could also be selected (not shown). The DHP-resistance was stable under non selective conditions, which demonstrated that a genetical alteration rather than a physiological adaptation underlies the improved osmotolerance.

Conclusions

The differences in the utilization efficiency of amino acids and betaines in the four Azospirillum spp. extends our previous report (10) and correlates it with the osmoregulatory properties of these species. The efficient growth on glycine betaine and choline as sole carbon and nitrogen sources by A. lipoferum may be used as a valuable taxonomic character. The possible role of proline, betaine and choline as compatible solutes in A. brasilense and A. halopraeferens resembles features of osmoregulation in Enterobacteriaceae (16), Rhizobium meliloti (17) and other osmotolerant bacteria (3). The DHP-resistant mutants of A. brasilense, which have improved environmental fitness towards water stress, may be useful for inoculation trials at semiarid conditions (5).

Acknowledgements

I am grateful to Drs. J. Döbereiner (Brazil), F. Favilli (Italy) and B. Reinhold (Germany) for providing Azospirillum strains. This work was supported by the Deutsche Forschungsgemeinschaft (SFB 137). I thank Prof. Dr. H. Stolp for helpful discussions and support of research with Azospirillum and Mrs. H. Ramming for skilful technical assistance.

References

1. Yancey, P.H., Clark, M.E., Hand, S.C., Bowlus, R.D. and Somero, G.N. (1982). Science 217, 1214-1222
2. Le Rudulier, D., Strom, A.R., Dandekar, A.M., Smith, L.T. and

Valentine, R.C. (1984). Science 224, 1064-1068

3. Imhoff, J.F. (1986). FEMS Microbiol. Rev. 39, 57-66
4. Wyn Jones, R.G. and Storey, R. (1981). In Paleg, L.G. and Aspinall, D. (eds.) Physiology and biochemistry of drought resistance in plants. p. 171-204. Academic Press Australia, Sydney
5. Sarig, S., Kapulnik, Y., Nur I. and Okon, Y. (1984). Exp. Agric. 20, 59-66
6. Reinhold, B., Hurek, T., Fendrik, I., Pot, B., Gillis, M., Kersters, K., Thielemans, S. and De Ley, J. (1987). Int. J. Syst. Bacteriol. 37, 43-51
7. Reinhold, B., Hurek, T., Niemann, E.-G. and Fendrik, I. (1986). Appl. Environ. Microbiol. 52, 520-526
8. Malik, K.A., Zafar, Y. and Hussain, A. (1980). Biologica 26, 107-111
9. Hartmann, A., Fu, H. and Burris, R.H. (1986). J. Bacteriol. 165, 864-870
10. Hartmann, A., Fu, H., Song, S.-D. and Burris, R.H. (1985). In Klingmüller, W. (ed.) Azospirillum III: Genetics, Physiology, Ecology. Springer Verlag, Heidelberg, p. 116-125
11. Tarrand, J.J., Krieg, N.R. and Döbereiner, J. (1978). Can. J. Microbiol. 24, 967-980
12. Sandhu, G.R., Aslam, Z., Salim, M., Sattar, A., Qureshi, R.H., Ahmad, N. and Wyn Jones, R.G. (1981). Plant Cell Environ. 4, 177-181
13. Ken-Dror, S., Preger, R. and Avi-Dor, Y. (1986). FEMS Microbiol. Rev. 39, 115-120
14. Tokuda, H. and Unemoto, T. (1984). J. Biol. Chem. 259, 7785-7790
15. Csonka, L.N. (1981). Mol. Gen. Genet. 182, 82-86
16. Le Rudulier, D. and Bouillard, L. (1983). Appl. Environ. Microbiol. 46, 152-159
17. Bernard, T., Pocard, J.-A., Perroud, B. and Le Rudulier, D. (1986). Arch. Microbiol. 143, 359-364

CARBON SOURCES FOR THE IN-VIVO NITRATE REDUCTASE ACTIVITY IN AZOSPIRILLUM BRASILENSE CELLS

Lakshmi Sadasivan and Carlos A. Neyra

Department of Biochemistry and Microbiology, Cook College

Rutgers University, New Brunswick, NJ 08903

Summary: Carbon sources representative of various entry pathways of carbon metabolism viz. fructose, gluconate, glucose and TCA cycle intermediates were used as energy sources to study the invivo nitrate reductase activity in whole cells of A brasilense ATCC 29145. Enzyme activity was linear when using cells grown and immediately assayed using nutrient broth. Organic acids such as malate, succinate, pyruvate, and gluconate were all effective as sources of energy for the reduction of nitrate to nitrite under aerobic conditions. Among the carbohydrates, glucose and sucrose could not support nitrate reduction, while fructose was able to supply energy after a lag of 10-20 minutes. This lag was also observed when using organic acids.

Most interesting was the observation that A brasilense cells were able to metabolize and use α-Ketoglutarate (α-KG) as an external source of energy for nitrate reduction. The use of α-KG as energy source by the A brasilense cells was confirmed by the uptake studies using radioactively labelled C^{14} α-KG. The cells did not grow in semi-solid nitrogen-free medium containing α-KG as sole carbon source but limited growth in liquid medium was observed with fixed nitrogen sources.

Keywords: Nitrate, reduction, nitrate reductase activity, carbon metabolism, A. brasilense.

Introduction

Carbon metabolism and growth of A brasilense under nitrogen fixing and non-nitrogen fixing conditions (using ammonium salts as the nitrogen source) has been studied and reported (5,10,15,24). Growth of these bacteria under KNO_3 as the sole source of nitrogen and the influence of carbon sources representative of various entry pathways on nitrate reduction has received limited attention however.

In A brasilense both assimilatory and dissimilatory types of nitrate reduction have been reported (2,9,11,12,14). In assimilatory nitrate reduction, nitrate is reduced to nitrite by nitrate reductase, under aerobic conditions. The nitrite thus formed is further reduced to ammonia by nitrite reductase before being used up for cell growth. In dissimilatory nitrate reduction however, the reduction of nitrate to nitrite occurs under anaerobic conditions and the nitrite formed is further reduced to N_2 by several enzymes with the production of ATP (13,22). Assimilatory nitrate reductase normally requires NADH/NADPH as electron donors.

In A brasilense cells, most reported work on nitrate reductase has been made using whole cells in an invivo assay system (2,9,11,13,14) with some exceptions (Ray and Mishra, 1985). The invivo nitrate reductase assay depends

Azospirillum IV: Genetics, Physiology, Ecology
Edited by W. Klingmüller

on the availability of the endogenous NADH/NADPH. Thus, factors that may influence the detection and expression of nitrate reductase activity in an invivo assay system, are diverse. It may depend on the uptake of substrate, presence or absence of suitable energy sources, uptake and metabolism of the carbon source, etc. Furthermore, A brasilense cells are known to possess the enzyme complements necessary for the operation of the Krebs Cycle and the Embden-Meyerhof-Parnas (EMP) and Entner-Doudoroff (ED) oxidative pathways (5,10,24) all of which are capable of providing reduced pyridin nucleotides (NADH/NADPH) for the reduction of nitrate to nitrite.

In the case of several plant species, carbohydrates have been shown to regulate enzyme activities by acting as a source of energy, carbon or co-enzyme for induced synthesis or activation of the existing enzyme molecule (1, 16-18). Sugars in general have been shown to have specific effects on NR levels in different plants like glucose inducing NRA in maize (2), prevention of the decline in NR activity in excised bean seedlings by exogenous supply of glucose and fructose (6, 21-23).

In this communication we report the influence of various carbon sources on nitrate reduction as well as growth using nitrate as the nitrogen source in A brasilense.

Materials and Methods

Bacterial Strains: Azospirillum brasilense Sp7 (ATCC 29145) was used for this study. The strain was routinely subcultured on nutrient agar (Difco) slants and maintained at 30° C.

Growth Curve Experiments:

Inoculum for growth experiments was harvested from a log phase culture grown in nutrient broth (NB; Difco) by centrifugation at 5000 rpm for 10 min. at 4° C, washed three times in equal volumes of 100mM $K\text{-}PO_4$ buffer (pH 6.8) and inoculated into different growth test media to an optical density at 660nm of < 0.1. Various test media were (1) Medium A (Ref. 14) supplemented with 10mM malate +5mM NH_4NO_3; (2) Medium A with 10mM α-Ketoglutaric acid [α-KG] + 5mM NH_4Cl; (3) Medium A + α-KG + 5mM NH_4NO_3; (4) Medium A +α-KG + 5mM KNO_3. Optical Density (OD_{660}) of the culture growing on a 200 rpm gyrotary shaker at 35° C was measured at 2 hour intervals.

Nitrate Reductase Activity

NRA was measured according to the method described by Neyra and Van Berkum (14) with modifications. The assay medium or mixture was basically the medium A with no malic acid or KOH added. Various carbon sources at a concentration of 10mM each were individually added. The substrate (KNO_3 0.5mM) was included in the 5ml assay mixture just prior to the addition of bacterial suspension (100ul containing 200ugm protein). Nitrite concentration was determined after an interval of 10-20 minutes in aliquots (100µl) from the assay mixture, by adding to them 2ml of 1:1 (v/v) freshly mixed solution of 0.02% N-napthyl-ethylene diamine dihydrochloride and 1% suphanilamide in '.5N HCl. Volume was made up to 4ml and after 15 minutes, color was read at 54)nm in a Gilford spectrophotometer. Protein estimation of the alkali (1N NaOH) digested bacterial suspension at 100° C for 7 minutes was followed by the Lowry's method (8).

C^{14} α-KG Uptake Studies

Cells were grown to mid-log phase in NB or minimal medium supplemented with α-KG as a carbon source. These were then harvested, washed once with minimal medium, and suspended in the same medium to 0.5mg of cell protein per ml. The suspension was incubated in a shaker at 30° C for 30 minutes. Uptake of [C^{14}] α-KG (NEN) was measured according to Gardiol et. al. (Ref. 4). [U-C^{14}]α KG uptake was measured by adding 500ul of cell suspension (300ug protein) to 450ul of minimal medium A containing 2nmoles of C^{14} α-KG plus 100uMoles of cold α-KG. Uptake was measured in other uptake media where 0.5mM KNO_3 or NH_4Cl were addeed as the nitrogen source. 100ul samples were removed at intervals of 10 minutes and the reaction was stopped by dilution with 2.0ml of cold minimal medium. Samples were then immediately filtered in a 0.45um porosity membrane filter. Filters were washed once more in 2.0 ml of the same medium. They were then dried under a heat lamp and radioactivity was determined in the scintillation counter Model Beckman LS 5801.

Results and Discussion

A. brasilense cells are able to use atmospheric N_2, ammonium or nitrate as sole N-sources for growth. However the growth patterns differ according to the carbon source used (10,15,20). For instance, flocculation have been reported to occur intensively in A. brasilense cells grown on fructose, in particular (20). This bacterium is also known to be unable to grow on glucose orα-Ketoglutarate as sole carbon source but possesses the necessary enzymes to metabolize either glucose or α-KG (10). Martinez-Drets et al. (10) suggested that A. brasilense may be cryptic for uptake of glucose or α-KG.

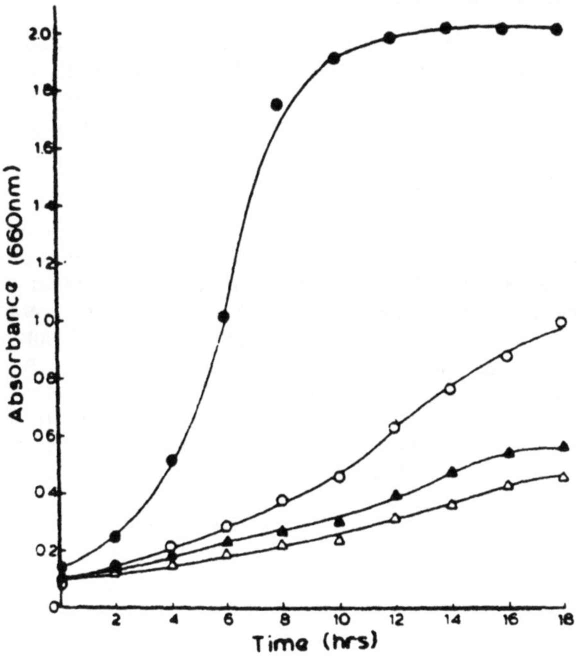

Figure 1. Growth pattern of A. brasilense ATCC 29145 in media containing different nitrogen sources plus α-Ketoglutaric acid (10mM). (O)NH_4Cl (△) KNO_3 (▲) NH_4NO_3; (●) malate + NH_4Cl

As shown in fig. 1, the capacity of A. brasilense cells to grow aerobically on α-KG media containing KNO_3 is also limited, but the cells could use α-KG as a sole carbon source for the reduction of NO_3^- to NO_2^- by whole cells (Table I). In fact, A. brasilense cells grown aerobically in nutrient broth, washed and suspended in NFb basal medium containing carbon sources representative of various entry pathways, viz. fructose, TCA cycle intermediates (malate, pyruvate, succinate or α-KG) and gluconate, exhibited average rates of invivo nitrate reductase activity between 4 to 5 umoles NO_2^- per hr per mg protein (Table I).

Table I. Invivo nitrate reductase activity by A. brasilense cells in assay mixtures containing different carbon sources[a].

Carbon Source	NR Activity
Minus Carbon	0
Sucrose	0
Glucose	0
Fructose	5.0
Pyruvate	3.5
Malate	4.3
α-Ketoglutarate	5.2
Succinate	5.5
Gluconate	4.2

a. The assay mixture contained Medium A with the carbon sources substituted by 10mM organic acids or sugars and 0.5mM KNO_3. The vials were incubated at 32° C for 1 hr and the NO_2^- produced was estimated as given in the Materials and Methods. Enzyme activities are expressed in umoles NO_2^- per hour per mg protein.

There was no measurable nitrate reductase activity in assay media containing either glucose or sucrose. The time course profiles for nitrate reductase activity using malate, fructose or α-KG as carbon sources were characterized by low initial rates (first 10 minutes) followed by linear rates for up to 30 minutes (figures 2 and 3). Calculations of enzyme rates within the linear portion gave maximal activities of up to 14 to 15 umoles NO_2^- per hr per mg protein. Most important was the observation that α-KG could support NR activity at rates comparable to carbon sources (fructose or malate) known to support good growth. The latter results could be interpreted as follows: 1) A. brasilense cells are not cryptic for α-KG uptake (see Figure 4) and 2) A. brasilense cells contain the necessary enzymes for α-KG metabolism and coupling to nitrate reduction. Alternatively, because α-KG can not support aerobic growth in media containing NO_3^- as sole N-source (figure 1), we could assume that α-KG may support dissimilatory NR activity at the membrane level. Assililatory NO_3^- reduction by A. brasilense has already been reported by Neyra and van Berkum (14) and Neyra et al. (13) among others (2,9,11,19). On the

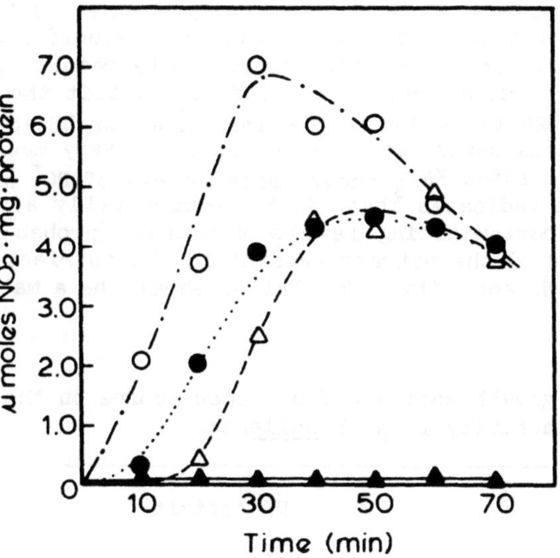

Figure 2. Time course of invivo NRA of A. brasilense in assay mixtures containing 0.5mM KNO_3 plus 10mM malate (●); 10mM fructose (△); 0.8% NB (○); and minus any carbon source (▲).

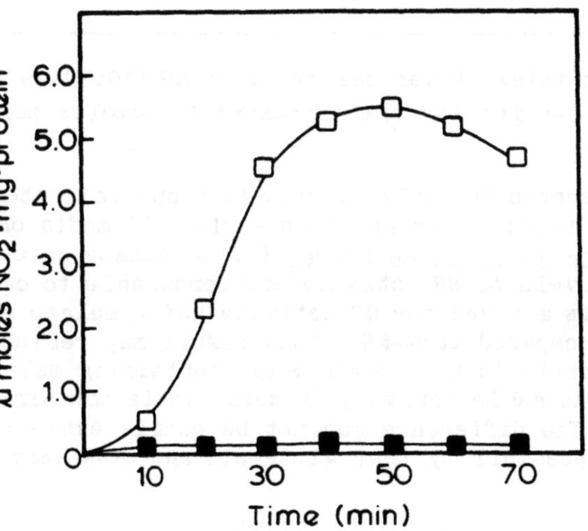

Figure 3. Time course of invivo NRA of A. brasilense in assay mixture containing 0.5mM KNO_3 plus 10mM glucose (■); and 10mM α-Ketoglutarate (□).

other hand, the initial low rates of NR activity observed using malate, fructose or α-KG may indicate the need for an activation of specific carriers for these substrates. This conclusion is further supported by results obtained using cells grown and assayed in nutrient broth which did not exhibit the initial lag (Figure 2). Even though the issue of NO_3^- induction for NR activity is not within the scope of this paper, we need to say that assay media containing only nutrient broth without added KNO_3 showed some levels of NO_2^- production (data not shown) which indicates that: A) the commercially available nutrient broth media contains contaminating levels of nitrate (probably from the beef extract component) and b) the nitrate present in the nutrient broth was sufficient to induce NR activity. The latter should be a matter of further investigation.

Table II. Effect of growth medium and nitrogen source on the invivo nitrate reductase activity in *A. brasilense*.

	NR Activity	
Growth Media	Malate[a]	α-KG
Nutrient Broth	7.41	5.5
Malate + NH_4NO_3	7.49	5.51
Malate + NH_4Cl	2.46	1.05
Malate + KNO_3	2.46	2.91
Glutamate	1.9	0

a. Assay mixture containing either malate or α-KG (10mM) as the carbon source plus 0.5 mM KNO_3. Activities are expressed in μmoles per hr per mg protein.

The results presented in Table II show that optimal rates of NR activity could be obtained by using cells grown on a rich, NB media or a defined NH_4NO_3 plus malate media. Cells grown on NH_4NO_3 (using malate as carbon source) showed the highest levels of NR activity and comparable to cells grown on nutrient broth. Cells assayed for NR activity using malate as carbon source showed higher rates compared to α-KG. This result may reflect the conditioning of the cells in the growth media containing malate. Cells grown on glutamate gave measurable activity in assay media containing malate but not in α-KG (Table II). The difference can not be easily explained at this time although there is a possibility that glutamate and α-KG may act through the GS/GOGAT system.

Addition to the assay mixture of several known metabolic inhibitors gave variable results (Table III).

Table III. Inhibitors of the invivo nitrate reductase activity in A. brasilense cells.

Inhibitors	NRA	% Inhibition Activity
Minus Inhibitors	3.36	0
Sodium Azide (1mM)	0.9	73
KCN (1 mM)	0.09	97
Chloramphenicol (50 ug/ml)	1.77	48
Cycloheximide (2 ug/ml)	1.15	66
Glutamine (100mM)	4.0	0
$(NH_4)_2SO_4$ (100mM)	1.68	50

a. NB grown cells were washed twice in minus carbon medium and alquots of 200 μl cells were resuspended in an assay medium containing 10mMα-KG, 0.5mM KNO_3, and various inhibitors listed in the table. Activities are expressed in μmoles per hour per mg protein.

Potassium cyanide and sodium azide known to inhibit respiration almost completely inhibited NR activity. This result could indicate that aerobic respiration is coupled to NO_3^- reduction and that cytochromes are involved in the electron transport but a more direct inhibitory effect on the NR enzyme per se cannot be disregarded. Similarly, the inhibition of NR activity by chloramphenicol and cycloheximide may not be specific, since both of these compounds are known to affect different cell functions. NR activity was inhibited but not completely (50% inhibition) by high levels of $(NH_4)_2SO_4$ at a concentration (100mM) known to inhibit nitrogenase activity. Addition of similar levels of Glutamine had no effect on the invivo NR activity indicating that direct regulation of enzyme activity by reduced-N may depend on the kind of reduced-N compounds present.

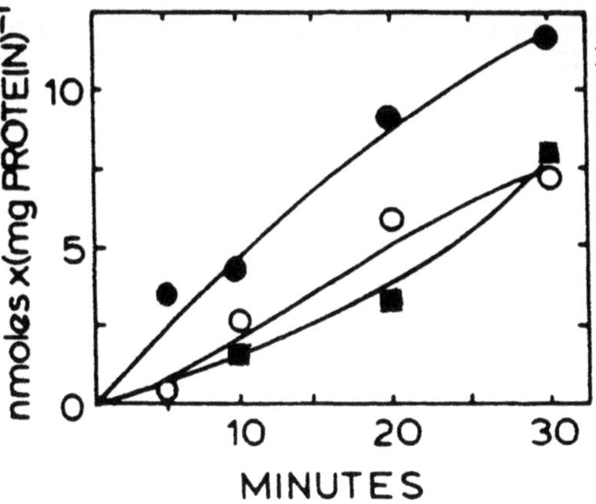

Figure 4. Uptake of (C^{14}) labelled α-Ketoglutarate by A. brasilene ATCC 29145. The cells were grown in nutrient broth. Symbols indicate presence of different nitrogen forms in the uptake medium. KNO_3 (○), plus NH_4Cl (●), or control minus N (■).

Conclusions

The characterization of the invivo NR activity in A. brasilense included the use of carbon sources in the assay medium with the purpose of 1) optimizing the invivo assay for NR activity in A. brasilense and 2) to study the relationships between NR activity and growth responses to carbon sources and their respective oxidative pathways. One of the main conclusions of this paper is that selectivity for carbon sources to support assimilatory growth on NO_3^- by A. brasilense is similar to the pattern of carbon utilization to support growth on NH_4^+ or N_2 (5,10,15,24). Aerobic growth on NO_3^- was supported by fructose, TCA cycle intermediates (pyruvate, malate, succinate) and gluconate but was not supported by glucose, sucrose or α-Ketoglutarate.

A. brasilense cells are known to possess the enzyme complements necessary for the oxidation of fructose (EMP pathway), organic acids (Kreb's cycle) and gluconate (ED pathway), (5,10,24). All of these three pathways appeared to be related in the supply of reducing power for nitrate reduction (Table I) but the exact metabolic complementation is not quite evident, at present. Further studies are needed to elucidate this point. More difficult to explain were the results obtained with α-KG. This organic acid does not support growth on either N_2, NH_4^+, or NO_3^-. However, the rates of NR activity using α-KG as a carbon source were comparable to fructose, malate, succinate, pyruvate, and gluconate known to support aerobic growth of A. brasilense. Furthermore, uptake of α-KG and the activity was inhibited by known respiratory inhibitors like cyanide and azide. These results could lead to the interpretation that α-KG supports only dissimilatory NO_3^- reduction. However, further research is needed to elucidate the apparent controversy created by the fact that α-KG does not support aerobic growth of A. brasilense while being able to support NR activity.

Acknowledgements

This work is published under the New Jersey Agricultural Experiment Station, Publication No. 0-01204-01-87, supported by State funds and United States hatch Act funds. We acknowledge the financial support to L.S. as Post Doctoral Fellowship by the NJAES, Department of Biochemistry and Microbiology and the International Agriculture and Food Program at Cook College, Rutgers University. We thank Dr. R.H. Burris, University of Wisconsin, Madison, for kindly supplying the radioactive C^{14} α-KG.

References

1. Aslam, M. and Oaks, A. (1975). Plant Physiol. 56,634-639.

2. Bothe, H., Klein, B., Stephan, M.P., and Dobereiner, J. (1981). Arch. Microbiol. 130,96-100.

3. Burke, K.A. and Lascelles, J. (1975). J. Bacteriol. 123,308-316.

4. Gardiol, A., Arias, A., Cervenansky, C., Gaggero, G., and Martinez-Drets, G. (1980). J. Bacteriol. 144,12-16.

5. Goebel, E.M., and Kreig, N.R. (1984). J. Bacteriol. 159,86-92.

6. Hanisch Jen Cate, C.H. and Bretler, H. (1981). Physiol. Plant. 52,129-135.

7. Hartman, A., Fu, H.-A., Song, S.-D., and Burris, R.H. (1985). In: Klingmuller, W. (ed.): Azospirillum III: Genetics, Physiology Ecology Springer Verlag, Heidelberg, p. 116-126.

8. Lowry, O.H., Rosenbrough, N.J., Farr, A.L., and Randall, R.J. (1951). J. Biol. chem. 193,265-275.

9. Magalhaes, L.M.S., Neyra, V.A.,and Dobereiner, J. (1978). Arch. Microbiol. 117,247-252.

10. Martinez-Drets, G., del Gallo, M., Burpee, C., and Burris, R.H. (1984). J. Bacteriol. 159,80-85.

11. Nelson, L.M. and Knowles, R. (1978). Can. J. Microbiol. 24,1395-1403.

12. Neuer, G., Kronenberg, A., and Bothe, H. (1985). Arch. Microbiol. 141,364-370.

13. Neyra, C.A., Dobereiner, J., Lalande, R.,and Knowles, R. (1977). Can. J. Microbiol. 23,300-305.

14. Neyra, C.A. and Van Berkum, P. (1977). Can. J. Microbiol. 23,306-310.

15. Okon, Y.S., Albrecht, S.L., and Burris, R.H. (1976). J. Bacteriol. 128,592-597.

16. Penteado, S.M., Zimmer, W., and Bothe, H.. (1984). Arch. Microbiol. 138,212-216.

17. Puranik, R.M. and Srivastava, H.S. (1983). Biochem. Physiol. Pflanz. 178,131-135.

18. Puranik, R.M. and Srivastava, H.S. (1983). Biochem. 22,2383-2387.

19. Ray Dutta, A.K. and Mishra, A.K. (1985). Curr. Sci. 54,637-640.

20. Sadasivan, L., and Neyra, C.A. (1985). J. Bacteriol. 163,716-723.

22. Srivastava, H.S. (1974). Indian J. Biochem. Biophys. 11,230-232.

23. Srivastava, H.S. (1980). Phytochemistry. 19,725-733.

24. Westby, S.A., Cutshall, D.S., and Vigil, G.V. (1983). J. Bacteriol. 156,1369-1372.

AN INFLUENCE OF PLANT GROWTH SUBSTANCES ON GROWTH AND NITROGENASE ACTIVITY FROM AZOSPIRILLUM BRASILIENSE

C. Christiansen-Weniger
Research Institute ITAL,
P.O. Box 48, 6700 AA Wageningen, The Netherlands

Summary: Plant growth substances, such as auxins, cytocinin, gibberillin and abciscin were tested for their effect on growth and nitrogenase activity from batch culture grown *Azospirillum brasilense*. The present results indicate that Indol-3-acetic acid (IAA) and 6-benzyl-aminopurine (BAP) can have a stimulating influence on the bacterial growth. Treatments with gibberillin (GA) and IAA in a final concentration of 2,5 mg/l are highly increasing the nitrogenase activity, while GA in a low concentration (0,025 mg/l) represses N_2 fixation. Comparing IAA, 2,4 D and 3,5 D with each other it could be demonstrated, that both, IAA and 2,4 D are positively effecting the nitrogen fixation, but 3,5 D does not. When tested for their effect on the NH_4^+-repression of nitrogenase it is only GA with a slight derepressing effect and this only in its lowest concentration tested. This effect became stronger and statistical more serious, when the GA addition was combined with an abscisic acid (ABA) or IAA treatment. ABA itself did not show any effect.
In conclusion this paper will demonstrate, that nitrogen fixation of the rhizosphere bacteria *Azospirillum* could be influenced by plant produced growth substances, what may be at least important in the discussion over the kind of association that exists between gramineae plant and *Azospirillum*.
Keywords: *Azospirillum*, plant growth substances, indol-3-acetic acid, 6-benzyl-aminopurine, gibberillin, abscisic-acid, 2,4 D.

Introduction

Involved in the discussion over the association between grammineae plant and *Azospirillum* spp. there are several works, in which the development of

Azospirillum IV: Genetics, Physiology, Ecology
Edited by W. Klingmüller

the nitrogenase activity in the root is observed over a longer period (1, 2, 3). It is remarkable, that higher activities often occur only in a late phase of plant-growth, shortly before flowering or even later. This could be effected by different factors, so for example by a decrease of nitrogenase repressing NH_4^+ in the rhizosphere, or in the same way, by a change of the O_2 tension. The knowledge here about is incomplete, because until yet it is not possible to measure the root bound acetylene reduction under controlled soil conditions. But even if the O_2 partial pressure in the root environment was hold to a constant level, nitrogenase activity increases drastically during panicle emergence (C. Christansen-Weniger, in preparation). Facing this, there was the idea, that the hostplant could influence the activity of the associated bacteria itself with help of its own hormonal regulatory system. To prove this some natural and artificial plant growth substances were tested for their effect on growth, nitrogenase activity and NH_4 nitrogenase repression of Azospirillum brasilense.

Material and Methods

The bacteria used in all experiments was Azospirillum brasilense (Wa 5), a strain isolated from the rhizosphere of a greenhouse grown spring wheat. For cell division studies the bacteria was grown in a batch on NB (28 °C), measuring the optical density at 450 nm. The plant growth substances were added in the begin of the experiment by sterile filtration in a final concentration of 2,5 mg/l, 4 replications each.

To measure the influence of the tested substances on the nitrogen fixation and further on the NH_4^+ nitrogenase repression the method described by Hartmann et al. was used (4). Azospirillum was grown in a batch with a controlled atmosphere of 88% Argon, 10% C_2H_2 and 2% O_2. The experiment was started by distributing 4 ml samples of this batch to anaerobe 39 ml serum-bottles, adding 1,5% O_2 and 10% C_2H_2 and shaking intensively at 28 °C. At 60 minutes the plant growth substances were injected. To induce nitrogenase repression, the culture was treated at 90 minutes with 0,8 mM NH_4Cl. An example for the development of acetylene reduction in such a system is given in Figure 1.

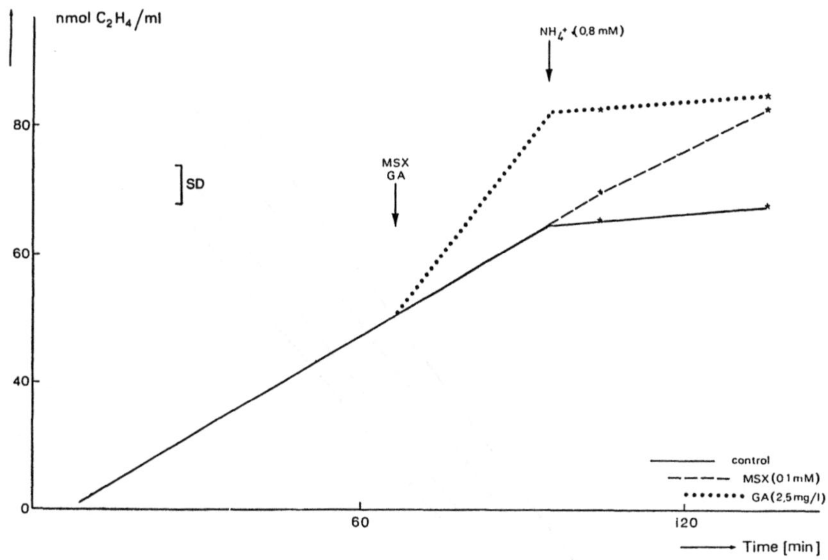

Figure 1. Acetylene reduction of an Azospirillum culture at 1,5% O_2 treated with MSX (0,1 mM) and GA (2,5 mg/1).

Because of the relative high experimental error involved in acetylene reduction assay all treatments were done with 6 replications each. Following substances were used:

Auxins: Indole-3-acetic acid (IAA), 2,4 dichlorphenoxy-acetic-acid (2,4 D), 3,5 dichlorphenoxy-acetic-acid (3,5 D) [1]

Cytocinin: 6-benzyl-aminopurine (BAP)

Gibberillin: gibberellic acid (GA)

Abscisin: abscisic acid (ABA)

The acidity of the stock solution was set upon pH 7 with HCl or NaOH.

Results and discussion

The influence of IAA, BAP and GA on growth of Azospirillum brasilense is demonstrated in figure 2. IAA and BAP showed both an increasing effect on bacterial growth. GA depressed cell division slightly but within the experimental standard difference. ABA (2,5 mg/1) did not influence the growth of the bacterial culture.

---

[1] (A gift from Dr. E. Knecht, Department of Plant-Physiology, Wageningen)

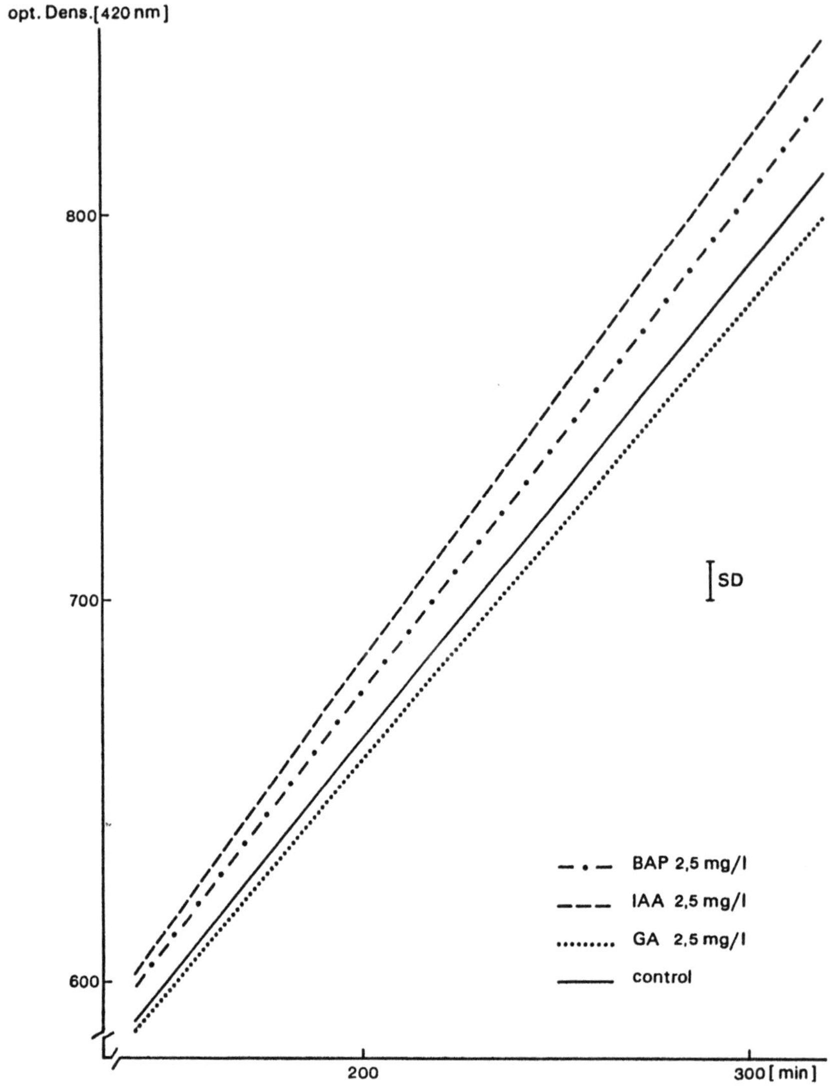

Figure 2. Influence of IAA, BAP and GA on growth of Azospirillum brasilense.

The influence of plant growth substances on the nitrogenase activity are given in Table 1. From the substances tested there were IAA and GA at a final concentration of 2,5 mg/l with a high increasing effect on nitrogen fixation. GA at a low concentration of 0,025 mg/l inhibited nitrogenase activity. ABA and BAP did not show an effect on the N_2 fixation of Azospirillum.

Table 1. Acetylene reduction of Azospirillum brasilense treated with plant growing substances at different final concentrations (nmol C_2H_4/h per ml bacterial suspension).

	0 mg/1	0,025 mg/1	0,25 mg/1	2,5 mg/1	25 mg/1
IAA	93	89	131	216	195
	b	b	ab	a	a
GA	93	71	115	126	107
	bc	c	ab	a	ab
BAP	93	73	82	92	84
	a	a	a	a	a
ABA	93	96	76	94	102
	a	a	a	a	a

Means with the same letter are not significantly different (p = 0,05).

To prove, whether the measured influence on nitrogen fixation is due to the regulatory ability of plant growth substances IAA was tested against other substances from the auxine group, such as 2,4 D and 3,5 D. 3,5 D is of nearly the same chemical structure as 2,4 D (Fig. 3), but without a comparable function in plant physiology (E. Knecht, pers. communication). As demonstrated in Table 2. IAA and 2,4 D induced a high increase of nitrogenase activity, but 3,5 D did not.

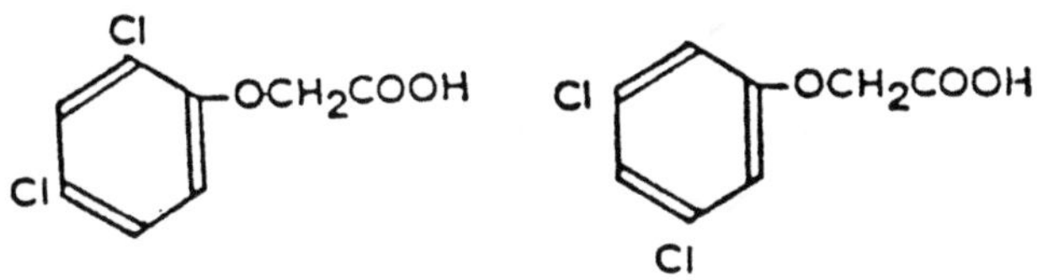

2,4 D 3,5 D

Figure 3.

Table 2. Acetylene reduction of Azospirillum brasilense treated with IAA, 2,4 D and 3,5 D at a final concentration of 2,5 mg/l (nmol C_2H_2/h per ml bacterial suspension).

Control	IAA	2,4 D	3,5 D
93	233	276	51
b	a	a	b

Means with the same letter are not significantly different (P = 0,05).

Testing the influence on the NH_4 nitrogenase repression only GA in a low final concentration of 0,025 mg/l was found to have a slight nitrogenase derepressing ability (Fig. 4). This effect was stimulated and became statistical more relevant when the GA treatment was combined with an addition of 2,5 mg/l ABA (Fig. 4). In the same way GA showed a higher nitrogenase derepressing activity together with IAA, 2,5 mg/l (Fig. 5). This derepression of nitrogen fixation effected by GA and ABA (IAA) may explain the fact, that acetylene reduction often is detected in Azospirillum inoculated roots even if the plant is treated with nitorgen fertilizer.

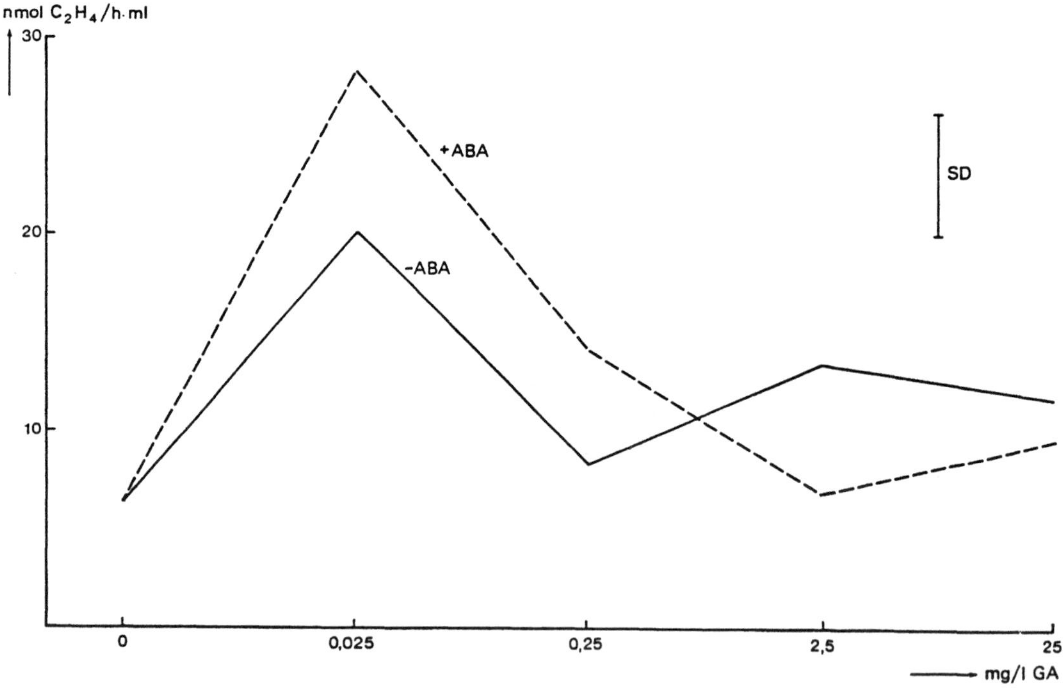

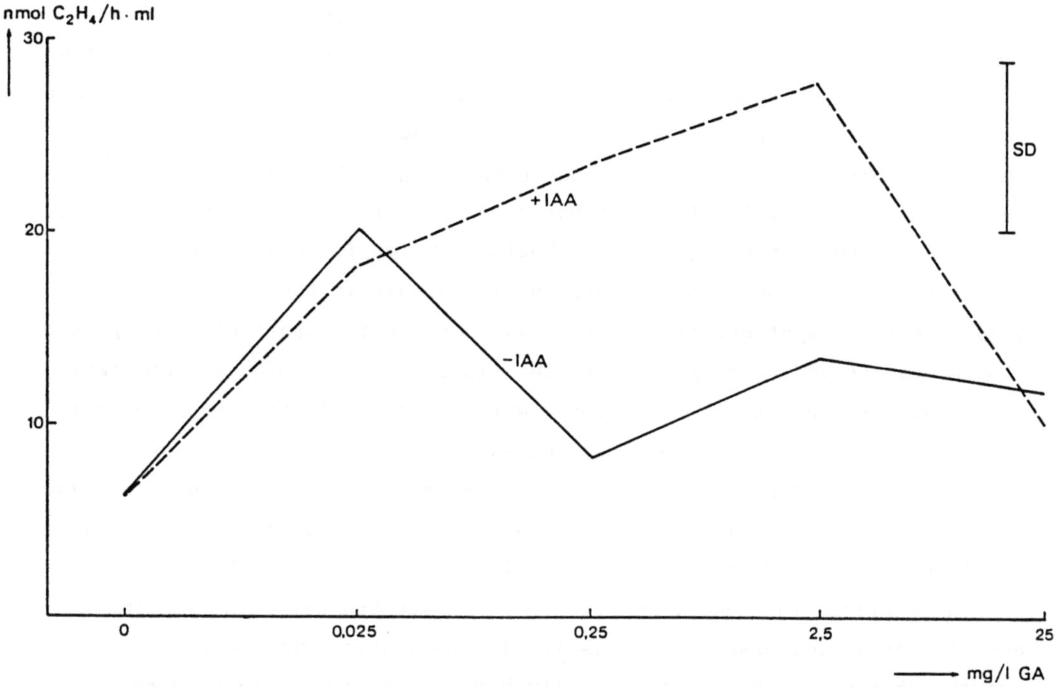

Figure 4 and 5. Effect of GA on NH_4^+ (0,8 mM) repressed nitrogenase from Azospirillum brasilense with and without ABA, 2,5 mg/l (Fig. 4) or IAA, 2,5 mg/l (Fig. 5).

How the plant-growth substances are interacting with the nitrogenase regulation of Azospirillum is not clear. There are several hints, which may indicate, that the here described influences are really hormonal effected, in some way comparable with the manner in which plant growth substances are regulating the physiology of plants: So, for example that 2,4 D increases nitrogenase activity, but not the nearly identic 3,5 D. Or the fact, that GA in a very low concentration shows a significant effect on nitrogen fixation, that disappears in treatments of higher concentration. Also the additive crossreaction between some growth substances so as described in figure 4. and 5. will lead to the same conclusion. But at least nearly nothing is known about the physiological way in which plant growth substances are effecting Azospirillum. It is obvious, that auxins are shown to be produced by Azospirillum itself (13), what may indicate, that it is possibly involved in bacterial physiology. Herefor further investigations have to be done.

For the ecology of Azospirillum-root interaction the here presented results may give a new and very interesting aspect. Specially the positive effect on nitrogenase and nitrogenase repression due to an IAA and GA treatment will be of great importance. Both plantgrowth substances are known to be involved in the regulation of root development. They could either be produced in the root itself or transferred there from the shoot (5, 6, 7). The same is true for ABA (9, 11). Cytocinin, which is shown here to have a positive influence on bacterial growth also occurs in the root and it is suggested to be synthetisized there (14). For the transport of plant growth substances inside the root the cortical tissue seems to play an important role (8). Moreover, there are several works which indicate, that the root tip is containing a high amount of auxins. (10, 11). So at least the root region, which is demonstrated in electron micrographs to be predestinately colonized by Azospirillum (12) is a place of possible availability of plant growth substances. Therefor there is evidence to conclude, that the Gramineae plant, in theory, is able to stimulate the nitorgen fixation activity of a root associated Azospirillum population with help of its growth substances. Whether this really happens in vivo would be very difficult to prove because of the strict effect plant growth substances always have on the plant.

Acknowledgements

I thank J.H. Becking and E. Knecht for helpful discussion, H. Beyer and W. van Lienden for technical help and M.J. 't Hart-Versteeg for typing this manuscript.

References

1. Wani,S.P. Upadhayaya,U.N. and Dart,P.J.(1984). Plant and Soil 82,15-29
2. Watanabe, J.and Chang Lin. (1984). Soil Sci. Plant Nutr.30,117-124
3. Christiansen-Weniger,C.Boddey,R.U. and Döbereiner,J.(1985). In Klingmüller,W.(ed.) Azospirillum III:Genetics,Physiology,Ecology. Springer-Verlag, Heidelberg, p.180-188
4. Hartmann,A.,Fu,H., Song,S. and Burris,R.H. (1985). In:Klingmüller,W.(ed.) Azospirillum III:Genetics,Physiology,Ecology Springer-Verlag,Heidelberg, p.116-126

5. Elliott,U.C.and Greenwood,U.S. (1974). Phytochemistry 13, 239-241

6. Scott,T.K. (1972). Annu. Rev. Plant. Physiol. 23, 235-258

7. Bridges,C., Hillmann,J.R. and Wilkins,U.B. (1973). Planta 115, 189-192

8. Shaw,S. and Wilkins,U.B. (1974). J. Exp. Bot. 25, 199-207

9. Hartung,W. and Behl,R. (1974). Planta 120, 299-305

10. Bernet,J.J. and Pillet,P.E. (1976). Planta 128, 183-184

11. Pillet,P.E. (1976). Planta 130, 245-249

12. Murty,U.G. and Saaka,J.K. (1987). Biol. Fertil. Soils 4, 3-7

13. Tien,T.U., Gaskins,U.H. and Hubbell,D.H. (1979). Appl. Envir. Microbiol. 37 (5), 1016-1024

14. Wareing, P.F., Horan, R., Henson, I.E. and Davis, W. (1977). In: Pilet, P.E. (ed.) Plant growth regulation. Springer-Verlag, Heidelberg, p. 147-153

INFLUENCE OF HERBICIDES ON GROWTH AND NITROGENASE ACTIVITY OF AZOSPRIILLUM

D. GADKARI
Lehrstuhl für Mikrobiologie , Universität Bayreuth
Postfach 101251, D-8580 Bayreuth, F.R.G.

Summary: The influence of the herbicides metamitron, metribuzin, ethiozin and paraquat on growth and nitrogenase activity of *Azospirillum lipoferum* and *A. brasilense* was studied. It was observed that metamitron (35 ppm and 70 ppm) and ethiozin (20 ppm) do not exert any inhibitory effect on nitrogenase activity of these strains. On the other hand, metribuzin (7 ppm and 14 ppm) and ethiozin (50 ppm) caused a marked decrease in nitrogenase activity. Additionally, the effect of ethiozin was studied under the following modes of application: A) ethiozin was added to the media 24 h after inoculation. B) Carbon source was added to the media 24 h after inoculation. According to the mode of treatment ethiozin caused different effects on nitrogenase activity. The herbicide paraquat (10 ppm) exerted a transient inhibitory effect on the growth. The growth of *A. brasilense* and *A. lipoferum* was inhibited for 12 h and 24 h, respectively. However, thereafter, growth and nitrogenase activity were completely recoverred. Key words: *A. lipoferum*; *A. brasilense*; growth; nitrogenase activity; herbicides; metamitron; metribuzin; ethiozin; paraquat.

Introduction

Azospirillum spp. are non-symbiotic nitrogen-fixing bacteria. They are widespread and common inhabitants of tropical and temperate soils (1; 2; 3). *Azospirillum* spp. are important members of the rhizosphere microflora. They proliferate in close vicinity of the roots, on the surface of the roots (4) or even in the intercellular system within the root tissue (2; 5). Their abili-

Azospirillum IV: Genetics, Physiology, Ecology
Edited by W. Klingmüller

ties to fix atmospheric nitrogen and to produce phytoharmones (6) are supposed to be beneficial to agricultural plants. Recently, additional useful effects of azospirilla causing increased plant growth (e.g. improvement of the water status of plants, root hair development, and increased mineral uptake by the roots) are described (7). During the past few years several workers reported that plants inoculated with Azospirillum spp. gave increased yields (e.g. 8; 9; 10; 11; 12).

In view of the importance of Azospirillum-plant associations it is necessary to know the influence of herbicides on Azospirillum spp. In modern agriculture, weed control by use of herbicides is a common practice. The herbicides and/or their metabolites do not only interfere with the growth of weeds but also interact with non-target soil microorganisms. It is, therefore, of interest to get a detailed knowledge of the side effects a particular herbicide exerts on microorganisms.

In the present work the influence of metamitron, metribuzin, ethiozin and paraquat on growth and dinitrogen fixation of Azospirillum lipoferum and Azospirillum brasilense was studied. In order to evade the complex situation in soil, the study was performed under laboratory conditions.

Material and Methods

Organism. Azospirillum lipoferum (ATCC 29708) and Azospirillum brasilense (ATCC 29145) were obtained from the American Type Culture Collection.

Media. For cultivation either tryptone water medium (Merck No. 10859) or semisolid nitrogen-free medium (13) were used.

Experimental procedure. From 48 h precultures (grown in nitrogen-free medium) 0.2 ml were inoculated into 30 ml Wheaton glas vials, each containing 10 ml of nitrogen-free semisolid medium. The vials were then incubated at 30°C. The cultures were examined daily for a period of 5 - 7 days with respect to N_2-ase activity.

Three sets of experiments were performed.

A) At the time of inoculation the carbon source (malate) and the herbicides were present in the media.

B) At the time of inoculation only the carbon source was present in the media. The herbicides were added after 24 h of incubation.

C) At the time of inoculation only the herbicides were present in the media. The carbon source was added after 24 h of incubation.

Growth. Growth was determined by dilution plate counts (CFU). 0.1 ml of appropriately diluted suspensions were spread on tryptone water plates (Trytone water + 1.0 % Agar).

Nitrogenase activity. Nitrogenase activity was measured by acetylene-reduction-test. The acetylene-reduction-test was performed as previously described (14).

Herbicides.

Goltix: Goltix is a commercial product available in powdary form. The active ingredient is metamitron (4,amino-3-methyl-6-phenyl-1,2,4-triazin-5-(4H)-one). The recommended field application ranges from 5 - 10 kg/ha.

Sencor: Sencor is a commercial product available in powdary form. The active ingredient is metribuzin (4,amino-6-tert-butyl-3-(methylthio)-1,2,4-triazin-5-(4H)-one). The recommended field application ranges from 0.5 - 3.0 kg/ha.

Tycor: Tycor is a new commercial product available in powdary form. The active ingredient is ethiozin (4-amino-6-tert-butyl-3-ethylthio-1,2,4-triazin-5-(4H)-one). The recommended field application ranges from 2 - 3 kg/ha.

These three preparations belong to the 1,2,4-triazin group. Commercial preparations contain 70 % of the active ingredient.

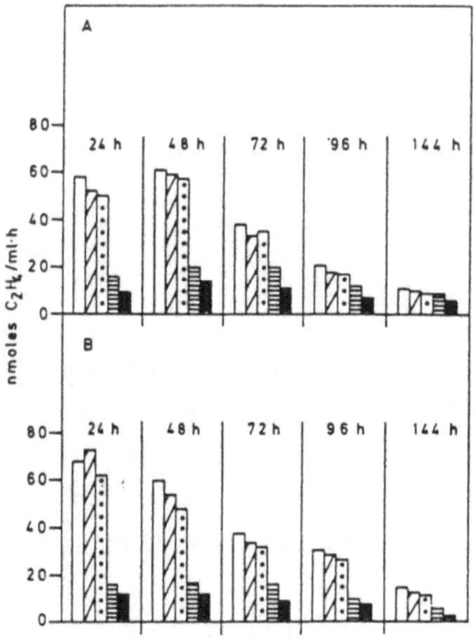

Fig. 1: Influence of metamitron and metribuzin on NA.
A: A. lipoferum.
B: A. brasilense.

- control
- 50 ppm metamitron
- 100 ppm metamitron
- 10 ppm metribuzin
- 20 ppm metribuzin

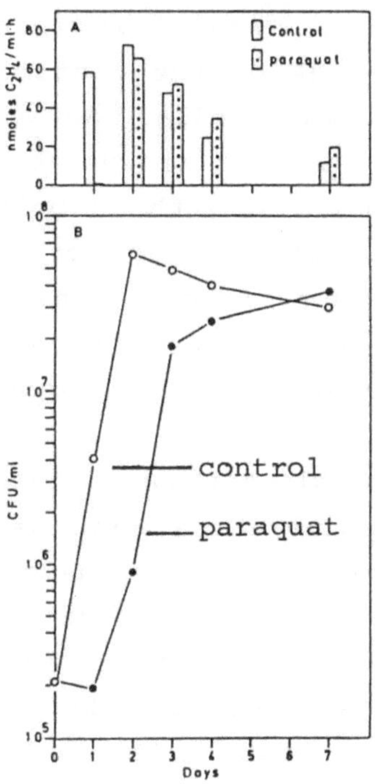

Fig. 2: Influence of 10 ppm paraquat on NA and growth of A. lipoferum. A: NA, B: growth.

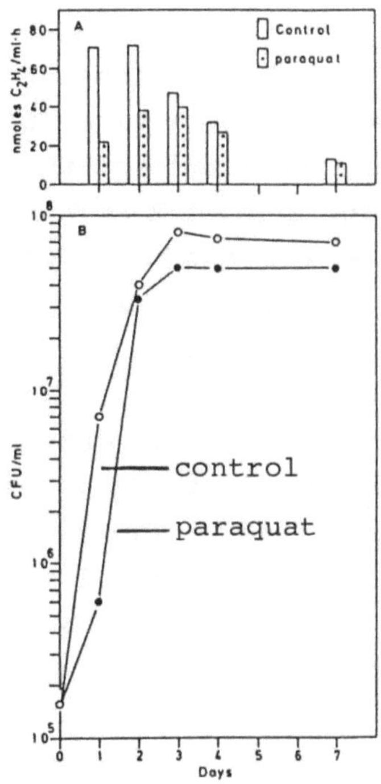

Fig. 3: Influence of 10 ppm paraquat on NA and growth of A. brasilense. A: NA, B: growth.

Gramoxone: Gramoxone is a commercial product available in liquid form. The active ingredient is paraquat (1,1-'dimethyl-4,4-bipyridium dichloride). Gramoxone contains 20 % of the active ingredient. The recommended field application ranges from 2 - 5 l/ha.

Results

In the presence of field rates of metamitron (35 ppm and 70 ppm) the nitrogenase activity (NA) of A. lipoferum and A. brasilense was not affected (Fig. 1 A, B). However, metribuzin (7 ppm and 14 ppm, field rates) caused an inhibition of NA. The inhibitory effect corresponded to metribuzin concentrations (Fig. 1 A, B).

In the presence of 10 ppm paraquat for the first 24 h the NA of A. lipoferum was completely inhibited (Fig. 2 A). The inhibitory effect was transient, after 48 h the NA of the control culture and the culture with 10 ppm paraquat was similar. Nevertheless, the growth of the culture with paraquat was considerably low (Fig. 2 B). After 7 days of incubation the growth in both cultures was similar. Compared to the control culture of A. brasilense, the culture in the presence of 10 ppm paraquat showed 70 % and 50 % inhibition of NA after 24 h and 48 h, respectively (Fig. 3 A). However, the growth in the control and paraquat culture was similar after 48 h (Fig. 3 B).

With regards to NA in the presence of ethiozin various results were obtained when the mode of the herbicide treatment was altered. At the time of inoculation when carbon source and herbicide ethiozin are present in the media, 10 ppm did not cause any inhibition of NA of A. lipoferum (Fig. 4 A). On the other hand, 30 % and 50 % inhibition was observed in the presence of 20 ppm and 50 ppm respectively. When ethiozin was given to the media after growth of the bacteria had already started (exponentially growing cells), the inhibitory effects (Fig. 4 B) of 20 ppm and 50 ppm were minor. Different pattern of NA of A. lipoferum were recorded when the cultures were first treated for 24 h with the

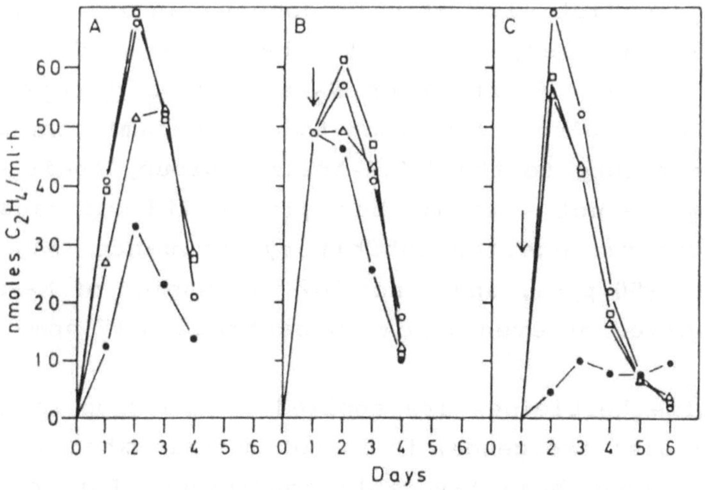

Fig.4: Influence of ethiozin on NA of A. lipoferum
A: At the time of inoculation malate and ethiozin were present in the media. B: At the time of inoculation only malate was present in the media. C: At the time of inoculation only herbicides were present in the media. ○——○, control; □——□, 10 ppm; △——△, 20 ppm; ●——● 50 ppm. ↓ addition of herbicides or malate.

herbicides, and the carbon source was later added to the media. Under these conditions 10 ppm and 20 ppm caused 20 % inhibition of NA. However, in the presence of 50 ppm the inhibitory effect was enhanced (Fig. 4 C). With different modes of ethiozin treatment, the herbicide effects on NA of A. brasilense (data not shown) were more or less similar to the effects observed for A. lipoferum.

Discussion

It is reported that herbicides, in general, promote the N_2-fixation of free-living diazotrophs (15). The herbicides dinoseb-acetate and ioxinil enhanced the N_2-ase activity of Spirillum lipoferum (16). It was also reported that the herbicide Alachlor caused a substancial increase in N_2-ase activity and this effect was explained as due to the increased population of A. lipoferum

(17). The N_2-ase activity of rhizosphere was stimulated due to Alachlor (19). However, none of the herbides used in this work caused any promoting effect on N_2-ase of A. lipoferum and A. brasilense. It is also clear, that though, metamitron, metribuzin and ethiozin belong to the 1,2,4-triazin group, their effects on Azospirillum are not similar. Metamitron did not cause any inhibition, ethiozin exerted inhibitory influence only at higher concentration (50 ppm), and a notable repression of N_2-ase activity was observed by even a low concentration (7 ppm) of metribuzin.

At the time herbicides are applied to the crop (e.g. preemergence and/or post emergence) it is not certain whether the rhizosphere diazotrophs have favorable conditions. During vegetation they may be subjected to carbon limitation and/or limitation of combined nitrogen. Obviously, the effects of herbicides on growth and N_2-fixation of the rhizosphere microorganisms vary. The results obtained with ethiozin showed quite clear that according to the mode of application the effects of herbicides differ. It appears that the susceptibility of bacteria towards herbicides is related to their physiological status.

It is reported that in the presence of 20 ppm paraquat the N_2-ase of Spirillum lipoferum was enhanced (16) for the first two days and thereafter, the N_2-ase activity was strongly inhibited. On the contrary, in the present study in the presence of 10 ppm paraquat N_2-ase of A. lipoferum was completely inhibited for the first 24 h. The N_2-ase of A. brasilense was for the first two days only 50 % of the control. In both Azospirillum cultures in the presence of paraquat the N_2-ase activity and increase in cell number did not correspond. The results obtained by use of paraquat revealed that the study of N_2-ase activity alone did not give proper informations of herbicide effects on growth of bacteria. Results presented in Fig. 2 and Fig. 3 exhibit clearly that a complete recovery of N_2-ase did not correspond to the growth and vice versa. Similar types of herbicide effects were observed for Azotobacter spp. (14).

Extrapolation of the effects observed in culture solutions to field conditions demands caution. Laboratory studies give only an idea of the possible effects. Therefore, further studies under field conditions are necessary.

References

1. Dommergues, Y., Balandreau, J., Rinaudo, C. and Weinhord, P. (1973). Soil Biol. Biochem. 5, 83-89

2. Döbereiner, J. and Day, J.M. (1976). In: Proceedings of the 1st International Symposium of Nitrogen Fixation (W.E. Newton and C.Y. Nyman, eds) Washington State University Press, Pullman, pp 518-538

3. Van Berkum, P. and Bohlool, B.B. (1980). Microbiol. Rev. 44, 491-517

4. Schank, S.C., Smith, R.L., Weiser, G.C., Zuberer, D.A., Bouton, J.H., Quesenberry, K.H., Tyler, M.E., Milam, J.R. and Littell, R.C. (1979). Soil Biol. Biochem. 11, 287-295

5. Lakshmi, V., Satyanarayana Rao, A., Vijayalakshmi, M., Lakshmi Kumari, K., Tilak, K.V.B.R. and Suba Rao, N.S. (1977). Proc. Indian Acad. Sci. Sect. B 86, 397-404

6. Tien, T.M., Gaskins, M.H. and Hubbel, D.H. (1979). Appl. Environ. Microbiol. 37, 1016-1024

7. Okon, Y. and Kapulnik, Y. (1986). Plant Soil 90, 3-16

8. Rennie, R.J. (1980). Can. J. Bot. 58, 21-24

9. O'Hara, G.W., Davey, M.R. and Lukas, J.A. (1981). Can. J. Microbiol. 27, 871-877

10. Kapulnik, Y., Kigel, J., Okon, Y., Nur, I. and Henis, Y.

(1981). Plant Soil 61, 65-70

11. Sarig, S., Okon, Y. and Blum, A. (1985). In: Nitrogen Fixation Research Progress (H.J. Evans, P.J. Bottomley, W.E. Newton, eds). Martinus Nijhoff Publishers, Dordrecht, Boston, Lancaster, p 707

12. Sarig, S., Kapulnik, Y. and Okon, Y. (1986). Plant Soil 90, 335-342

13. Albrecht, S.L. and Okon, Y. (1980). In: Methods in Enzymology. Vol. 69 (A. San Pietro ed). Academic Press, London, New York, San Francisco, pp 740-749

14. Gadkari, D. (1987). Zentralbl. Mikrobiol. 142, 283-291

15. Anderson, J.R. (1978). In: Pesticide Microbiology (I.R. Hill and S.J.L. Wright, eds). Academic Press, London, New York, San Francisco, pp 313-533

16. Jagnow, G. Heinemeyer, O. and Draeger, S. (1979). Ecotox. Environm. Safety 3, 152-158

17. Marriel, I.E. and Cruz, J.C. (1978). In: Limitation and Potentials for Biological Nitrogen Fixation in the Tropics (J. Döbereiner, R.H. Burris, A. Hollaender, eds). Plenum Press, New York, London, p 340

18. Berggold, R., Krotzky, A., Jäger, D., Dart, P.J. and Werner, D. (1983). Z. Pflanzenernähr. Bodenk. 146, 643-652

THE EFFECT OF DICOFOL ON MORPHOLOGY GROWTH AND NITROGENASE ACTIVITY OF AZOSPIRILLUM LIPOFERUM

D.M.S. Mano, A.C.M. Matos and T. Langenbach

Universidade Federal do Rio de Janeiro, Instituto de Microbiologia Departamento de Microbiologia Geral, Ilha do Fundão, 21944 - Rio de Janeiro, RJ, Brazil.

Summary: Modifications of Azospirillum lipoferum Sp Br17 were observed when it was incubated with the organochloride insecticide Dicofol. In experiments run under non N_2 fixing conditions, cellular volume measured in a Coulter Counter Channalyser was reduced immediately after addition of pesticide. Cell counts, made in a Coulter Counter increased during 12 hours after addition of 250 and 500 ppm of Dicofol. Cell viability was reduced by 1 hour incubation in Dicofol concentrations of 10 to 500 ppm. These results suggested that one of the first consequences of Dicofol action was cell disruption. Incubation for 1 to 7 days showed that maximum cell counts, protein content, average cell volume, and viability were lowered by increased Dicofol concentrations. Growth was totally inhibited by 250 and 500 ppm of Dicofol but in some experiments a latent growth was observed, suggesting selection of resistant mutants. Experiments under N_2 fixing conditions showed that nitrogenase activity was inhibited by increasing Dicofol concentrations of up to 1000 ppm, but in some experiments was stimulated by a concentration of 1 ppm. Microscopic observations of Azospirillum morphology in the presence of Dicofol showed cyst like formations that may be an induced resistance mechanism.

Introduction

One of the biggest ecotoxicological problems in Brazilian agriculture is the intensive use of pesticides with mechanisms of action that can attack non target organisms, including soil microorganisms.

In vitro experiments have shown pesticides to act on microorganisms in many different ways including the reduction of their survival but in soils these products had little effect (1,2). This apparent contradiction between field and *in vitro* observations could be due to chemical process in soil (3) and possible induction of resistance or selection of resistant mutants to pesticides (4).

Azospirillum IV: Genetics, Physiology, Ecology
Edited by W. Klingmüller

Our objective was to study the effects of the organochloride insecticide Dicofol on the physiology and morphology of A. lipoferum, wich was choosen as a model because of its ecological and agricultural importance.

Material and Methods

Azospirillum lipoferum Sp Br17 (ATCC 29709) was received from J.Döbereiner (EMBRAPA, Km 47, Rio de Janeiro). Stock cultures were mantained in semi-solid potato medium (pH 6.8-7.0), after growth during 24 a 48 hours at 32ºC. The inoculum was grown overnight with agitation in NFB broth (5) enriched with NH_4NO_3, (0.8 g/l), at 32ºC. Cells were centrifuged, washed and resuspended in the same medium without NH_4NO_3. The final concentration of inoculum was about $2x10^5$ cell/ml under non N_2 fixing conditions. In experiments run under non N_2 fixing conditions, 100 ml of liquid NFB containing Dicofol in different concentrations, was distributed in 250 ml erlenmeyer flasks. The culture was incubated on a shaker at 32ºC. Aliquots were taken with 0, 0.5, 1, 2, 3 and 7 days of incubation. Growth was evaluated by protein content measured by Lowry Method (6), cell numbers by particle counts of Coulter Counter and volume by Coulter Counter Channalyser, and viable counts by the spread plate method.

Cell morphology observations were performed by optical microscopy with 400X magnification and by electron microscopy. In the latter cells were pre-washed in cacodylate buffer 0.2M, fixed with 2.5% glutaraldehyde, pos-fixed with osmium tetroxide and stained with uranyl acetate (30 min.) and lead citrate (5 min.) (7).

Experiments under N_2 fixing conditions were performed in semi-solid NFB medium distributed in penicillin flasks and incubated at 32ºC. Two sets of experiments were run in triplicate: in the first, Azospirillum were grown without Dicofol during 24 hours and then put in contact with different concentrations of Dicofol for 1 hour incubation at 32ºC with acetylene; in the second group of experiments, cultures were grown in the presence of Dicofol for 24 hours followed by 1 hour incubation at 32ºC with acetylene. Dicofol was added to each flask in 100 ul of different dilutions of stock solutions. Nitrogenase activity was determined by the acetylene reduction assay and in the same samples protein content was determined by Lowry et. al. (6)

Results

Growth under non fixing conditions:

Growth kinetics measured by protein content showed a lag phase during the first twelve hours of incubation, followed by logarithmic growth. Both growth rate and the maximum population values in the stationary phase were reduced proportionally to the amount of pesticide added (Fig. 1). The same pattern of kinetics was observed in Coulter Counter particle counts with concentrations below 100 ppm. At 250 and 500 ppm of Dicofol the particle counts increased by one order of magnitude during the first 12 hours, and decreased slowly during the next 7 days of incubation. An increase in counts was observed on the seventh day in some experiments (Fig. 2).

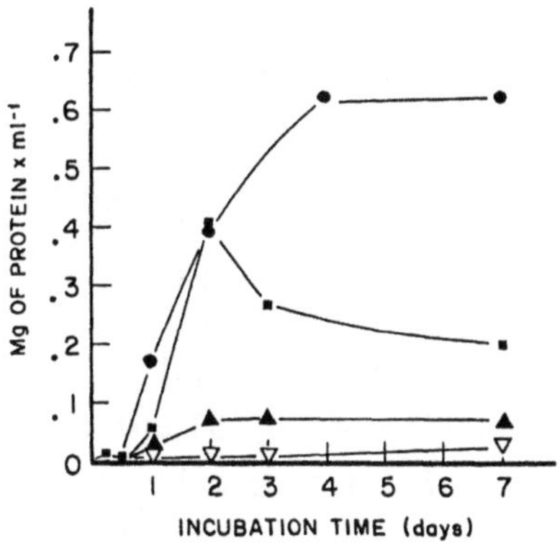

Fig. 1. Growth of A. lipoferum Sp Br17 in presence of Dicofol, measured by protein content (Lowry). (●) control, (▪)10 ppm, (▲)50 ppm and (▽)500 ppm.

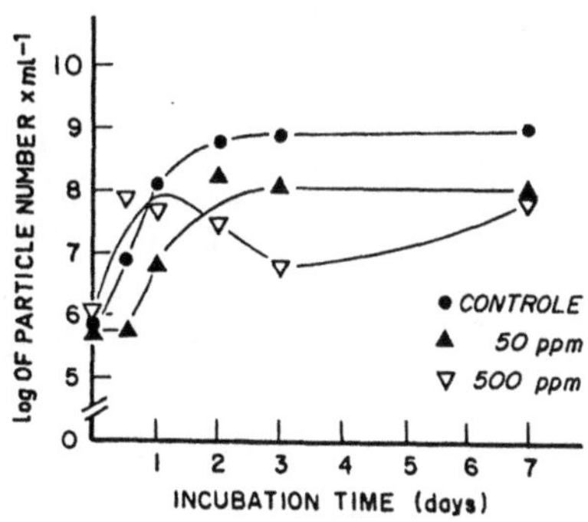

Fig. 2. Particle counts of A. lipoferum Sp Br17 in presence of Dicofol, measured by Coulter Counter. (●) control, (▲)50 ppm and (▽) 500 ppm.

Cell volume measurements with 1 to 10 ppm of Dicofol were similar to the control, but with concentrations of 50 ppm and higher an immediate reduction occured, followed by stable cell volume during the 7 days incubation period (Fig. 3). The histograms of cells volumes at 50, 100, 250 and 500 ppm of Dicofol were distinct from that of control populations, showing a similar distribution pattern to that of the background. Cell viability was reduced in proportion to the concentration of active ingredient of the pesticide during 1 hour of incubation with elevated Dicofol concentrations, and there were increases in growth rate after the first hour with concentrations up to 100 ppm of Dicofol. Maximum values of viable cells were reached after 48 hours of growth, and were reduced by increased Dicofol concentrations. Viability was reduced to near zero levels by 250 and 500 ppm of Dicofol but in some experiments samples with viable cells were found (Fig. 4).

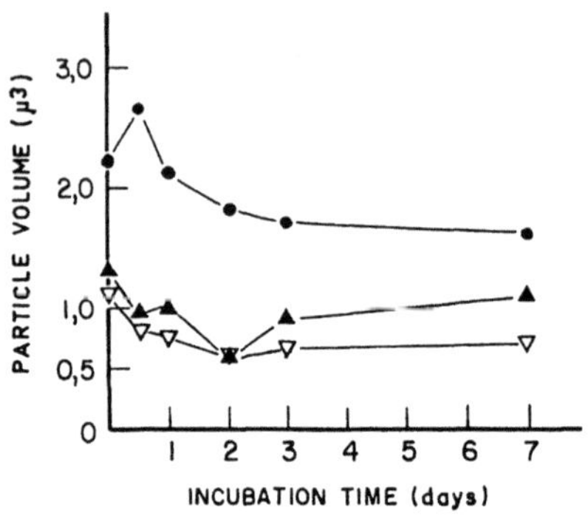

Fig. 3 Particle volume of A.lipoferum Sp Br17 cells incubated with of Dicofol. (●) control , (▲) 50 ppm and (▽) 500 ppm

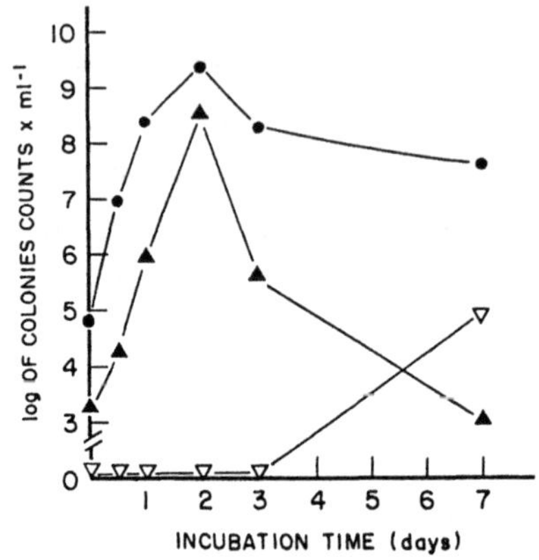

Fig. 4. Growth of A.lipoferum Sp Br17 in presence of Dicofol measured by viable counts. (●) control , (▲) 50 ppm and (▽) 500 ppm.

Absence of motility and shrinkage of cells was observed by light microscopy during 24-48 hours immediately after addition of 10, 50 and 100 ppm of Dicofol. In the presence of 250 to 500 ppm of Dicofol intact cells were not observed. Cysts began to appear after 24 to 48 hours incubation with 10 to 100 ppm of Dicofol or higher concentrations when latent growth occured. Giants cysts surrounded by an outer layer were observed in electron micrographs (Fig. 5).

Each cyst contained 1 to 4 central bodies with several electron transparents areas which were demonstrated to be poly-β-hydroxybutirate granules by light microscopy staining methods (Fig. 6).

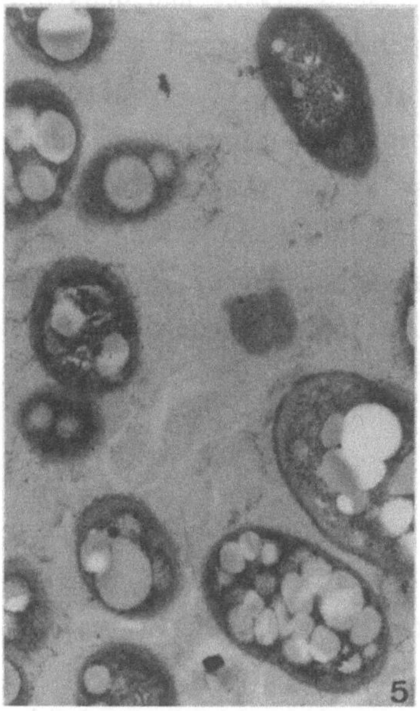

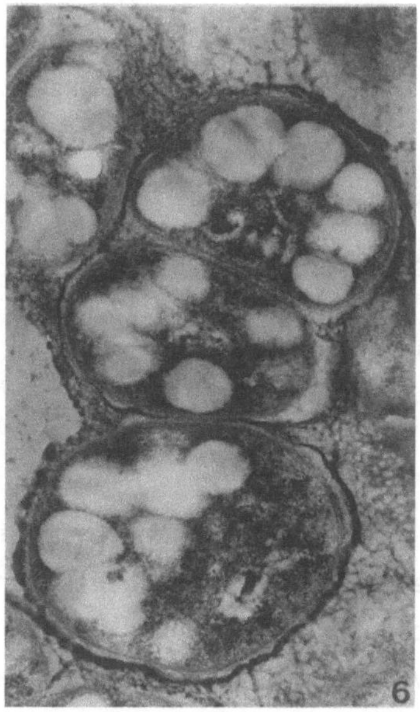

Fig. 5. Thin section of A. lipoferum cysts in different formations, after 48 hours incubation with 50 ppm of Dicofol.(X 13.500).

Fig. 6. Thin section of A. lipoferum cyst at higher magnification, showing 3 central bodies with PHB granules, surrounded by an outer layer (X 17.000).

Growth and nitrogenase activity under N_2 fixing conditions:

Growth patterns in the presence of Dicofol were similar under fixing and non fixing conditions during 24 hours incubation. The inhibition started with 1 ppm of Dicofol and increased gradually over a range of three orders of magnitude. In experiments with 1 and 24 hours incubation with Dicofol, nitrogenase activity patterns were similar to protein growth curves. A slight increase in growth and nitrogenase activity rather than inhibition was observed in some experiments at a concentration of 1 ppm (Fig. 7).

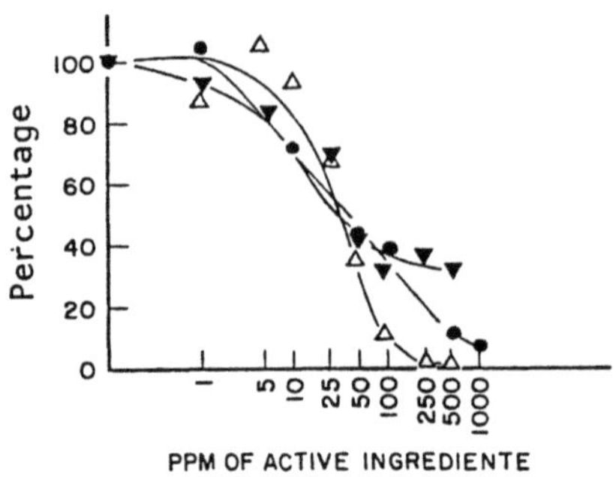

Fig. 7. Nitrogenase activity and growth with increasing Dicofol concentrations. (●) Percentage of N_2ase activity after 1 hour of incubation with Dicofol; (△) Percentage of N_2ase activity after 24 h of incubation with Dicofol and (▼) percentage of growth measured by protein content after 24 h incubation under non N_2 fixing conditions.

Discussion

Dicofol inhibited cellular growth rate, maximum cell growth and viable cell counts. The increase in particle counts during the first 12 h of incubation in high Dicofol conentrations could be due to cell disruption in smaller fragments since cell volume and viability were also reduced. The lipophilic caracter of Dicofol (8) and the known imbedment of organochloride insecticides in the lipid components of the membrane bilayer (9) suggest that the observed effects could have been the result of cytoplasma membrane disruption. Increasing growth and cell viability noted during some experiments with 500ppm Dicofol with 7 days incubation, suggest the appearance of resistant forms.

Nitrogenase activity and growth measured as protein had similar graphic patterns. Since nitrogenase activity was expressed relative to protein, the results with 24 h Dicofol incubation experiments shows that this product inhibited nitrogenase activity more than cell viability. Specific nitrogenase activity would have been held at a constant value if its inhibition had been only the result of cell lysis.

Morphology of the cyst formed, was similar to that observed by Sadavisan and Neyra (10) in A, brasilense growth under different cultural conditions. The cyst could be a protective mechanism against the presence of pesticides and other adverse environmental conditions. It is possible that this resistance is the result of cyst formation.

Acknowledgments

We thank Dr. Allen Norton Hagler and Dr. Leda Cristina Mendonça-Hagler for their help in the preparation of the manuscript and Dr. Wanderley de Souza for the E.M. plates. The work has been supported by grants from CNPq, FINEP and the Comission of the European Community.

References

1. Lal, R. and Saxena, D.M. (1982) Microbiol. Rev. 46, 95-127

2. Reddy, B.V.P., Dhanaraj, P.S. and Rao, V.V.S.N. (1985) in Lal, R. (ed.) Insecticide Microbiology. pp. 169-201

3. Lal, R. (1983) Crit. Rev. Microbiol. 10, 261-295

4. Lal, R. and Saxena, D.M. (1980) Residue Reviews 73, 49-85

5. Krieg, N.R. and Döbereiner, J. (1984).In Krieg, N.R. and Holt, J.G. (eds.): Bergey's Manual of Systematic Bacteriology, Williams and Wilkins, Baltimore, 1, 94-104

6. Lowry, O.M., Rosebrough, N.J., Farr, A.L. and Randall, R.J. (1951). J. Biol. Chem. 193, 265-275

7. Cole, R.M. and Popkin, T.J. (1981) Electron Microscopy. In Gerhardt, R.G., Murray, E., Costilow,R.N., Nester, E.W., Wood, W.A., Kreig, N.R. and Phillips G.B. (eds.), Manual of methods for general bacteriology. American Society for Microbiology, Washington, D.C. pp. 34-51

8. Brooks, G.T. (1974) Chlorinated Insecticides, Vol. I, II. CRC Press, Cleveland, Ohio.

9. Buff, K. and Berndt, J. (1981) B.B.A., 643, 205-212

10. Sadasivan, L. and Neyra, C. (1985) In Klingmüller, W. (ed.) Azospirillum III: Genetics, Physiology, Ecology. Springer-Verlag, Heidelberg, p. 230-242

INTERACTION BETWEEN AZOSPIRILLUM BRASILENSE CD AND WHEAT ROOT CELLS DURING EARLY STAGES OF ROOT COLONIZATION

Y. Bashan and H. Levanony

Department of Plant Genetics, The Weizmann Institute of Science, Rehovot 76 100, Israel

Summary: Following inoculation of wheat seedlings by A. brasilense Cd, the bacteria were strongly adsorbed to soil particles, but could migrate in the soil towards the plants. Upon reaching the wheat roots, the bacteria formed relatively small aggregates on the root surface, and a larger internal population within the cortex; no bacteria were found in the vascular system. Several bacteria penetrated into young live root cells, but most were free in the intercellular spaces or bound to cortical and epidermal cell walls by an electron dense material. The bacteria induced an increase in the rate of proton extrusion of the roots as well as an increase in cell division of the root tip.

Introduction

Inoculation of plants with bacteria of the genus Azospirillum has been tested world-wide in view of their potential contribution to plant productivity (1,2). This practice suffers from several disadvantages, of which the most notable ones are the inconsistent and unpredictable yield results, i.e., significant increases in yield besides negligible or even reduction in yield (3), and a high percentage of roots which fail to be colonized by the applied bacteria (4). Several factors, including the limited migration of the bacteria in the soil, adsorption of bacteria to soil particles and competition with the natural rhizosphere bacteria for nutrients that leak from the roots decrease the size of Azospirillum population and in turn their potential contribution to plant growth (4,5). The obtained erratic responses to inoculation led Schank and Smith (3) to assume that the success of Azospirillum inoculation was short-lived and this technology should not yet be transferred to farmers. However, more positive data was accumulated since (4), and a considerable interest in Azospirillum still persists world-wide, despite its controversial nature.

At least three types of factors, different in nature, participate in the plant-bacteria association, (i) soil factors, including physical, chemical and biological variables, (ii) bacterial factors, including metabolism and survival of the applied bacteria as well as bacterial factors that affect plant growth, and (iii) plant factors including the capability to respond to the metabolic activities of the bacteria. All these factors operate simultaneously from time of inoculation - usually performed at sowing or at the seedling stage (6). Therefore, inoculating

Azospirillum IV: Genetics, Physiology, Ecology
Edited by W. Klingmüller

plants with Azospirillum on a large experimental scale should not be recommended before these factors are elucidated.

Since we assumed that the initial stages in the association between the two organisms are crucial for the success of inoculation, we focused on these stages and attempted to identify factors that hinder this association. Our model for this study consisted of the wheat plant as a host and A. brasilense Cd as the inoculum. This bacterial strain was selected because it has been relatively successful in Israel, U.S.A. as well as in other countries (7,8) and because it is one of the most common reference strains in the world.

Adsorption to soil particles

Adsorption of A. brasilense Cd to soil was estimated in γ-irradiated soil columns filled with various types of soil, after washings with sterile tap water. Enumeration of bacteria in the soil and in the eluate was done as previously described (9,10). Live and dead cells of A. brasilense Cd strongly adsorbed to light and heavy soil and only slightly to quartz sand. An increase in bacterial adsorption was obtained by increasing soil content with clays and organic matter, by a decrease in soil pH or by flooding the soil with water, whereas presence of bacterial attractants, an increase in soil pH and drying the soil decreased bacterial adsorption. Intensive washings of the soil with water did not detach the bacteria from soil particles but only from sand particles, which acted as a bacterial filter. Overwashing the soil recovered relatively few bacteria, whereas most of the bacterial population in sand was removed. Bacteria were adsorbed to the upper layer of the soil profile but infiltrated deeper if the soil was very dry.

We propose that most of the population of A. brasilense Cd is adsorbed to soil particles immediately after inoculation and does not penetrate to deeper soil layers even after intensive water application. Bacterial adsorption counteracted bacterial motility. Factors affecting motility are described below.

Self motility and transport in the soil

A. brasilense Cd is known to be motile towards several synthetic attractants (11,12). We measured bacterial self motility in the soil and bacterial transfer in the field by methods described elsewhere (13,14). The main factor affecting motility towards germinating seedlings was soil moisture. The highest migration rate was obtained in brown-red degrading sandy soil at near field capacity (16% v/w water) and above. At 20% moisture, this soil tended to become flooded, facilitating motility of bacteria in the free water films even in the absence of seedlings (Fig. 1a). Of secondary importance were the soil type and the duration of plant growth prior to bacterial application: migration decreased with increasing soil density and increased when plants were grown in the soil several days (1-3 d) before inoculation. Migration was initiated following a lag period of 24-36h, and was characterized by a bacterial band migrating through the soil towards the target roots.

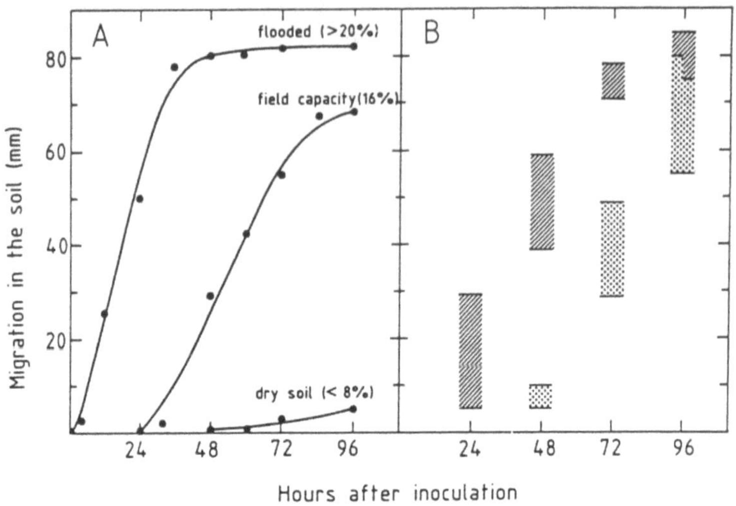

Fig. 1: Migration of A. brasilense Cd in the soil. A - in dry soil; soil in field capacity; and flooded soil. B - ▨ , Migration towards 11 wheat cultivars and , ▩ towards glycine and aspartic acid. Each column represents the maximal and minimal migration at the indicated hour after inoculation.

Eleven common and hard wheat cultivars were found to stimulate significantly bacterial migration. A similar effect was obtained by several bacterial attractants such as glycine and aspartic acid, known to comprise part of the grasses exudates, when encapsulated into solid alginate beads (15). These attractants, however, did not differ in the intensity of their effect (Fig. 1b). Therefore, we propose that A. brasilense Cd is stimulated to migrate non-specifically towards growing wheat plants.

Horizontal transport of A. brasilense Cd in the soil and its vertical transport along plant rhizosphere were measured. No transport was detected in the absence of living plants. Under controlled conditions the bacteria migrated horizontally at least 30 cm from the inoculation site towards the growing plant. Once the first root system was colonized, all the neighboring plants became colonized. Horizontal transport under field conditions was at least 160 cm, and depended on the presence of live plant roots. Weeds such as wild oat, wild barley, Phalaris paradoxa L., P. brachystachys Lk, Malva aegyptia L., Notobasis syriaca L. Cass.

and Silybum marianum L., that grew in the wheat field served as bacterial carriers. Vertical transport in soil columns under controlled environment, in the presence of a live plant was up to 40 cm. Under field conditions, bacteria were detected as deep as 50 cm, within the root systems of the plants. During the winter season A. brasilense Cd was found mostly in young roots at a depth of 20-50 cm and near soil surface. Profiles of depth distribution of A. brasilense Cd carried out at four different locations in Israel showed a non-homogeneous colonization pattern within the same root system or between adjacent plants.

We propose that A. brasilense Cd is transported in the soil horizontally and vertically and that these transports depend mainly on the presence of plants.

Adsorption on the root surface and on root hairs

Adsorption of Azospirillum cells to the root surfaces or to root hairs of several plant species were reported (16,17,18,19). When A. brasilense Cd was inoculated onto wheat roots or reached them by its own motility, the bacteria multiplied and formed an aggregate form of colonization. These bacterial cells were attached to each other and to the root surface by a fibrillar material (Fig. 2a, b). Slight rinsing of the roots (20) released most of this population. Disinfection of highly colonized roots eliminated most of the bacteria, whereas inoculating live or dead roots with γ-irradiated killed bacteria resulted in very small adsorption. However, killing roots before inoculation had no effect on bacterial adsorption to the roots. γ-irradiation following an intensive root colonization resulted in elimination of most of the dead bacterial cells from the root surfaces.

Microscopic inspection of external wheat roots surfaces revealed bacteria along the roots - mostly in the elongation zone. However, the root hairs were not heavily colonized. Small aggregates as well as single bacteria were found on root hairs. Few bacteria penetrated the root hairs, as observed by transmission electron microscopy. Usually, there were less bacteria on root hairs than on the root surface in the same region.

We proposed that active metabolism of A. brasilense Cd is essential for root colonization. Since this species is able to colonize and adsorb to dead roots, the active metabolism of the root probably does not play an important role in this association. Thus, the adsorption of A. brasilense Cd to the surfaces of wheat roots can be defined as a weak, metabolism-dependent process.

Internal root colonization

There were indications that Azospirillum species could penetrate the roots and be established in the intercellular spaces of their host plant (20,21). However, there is no solid evidence that the observed bacteria were the inoculated ones.

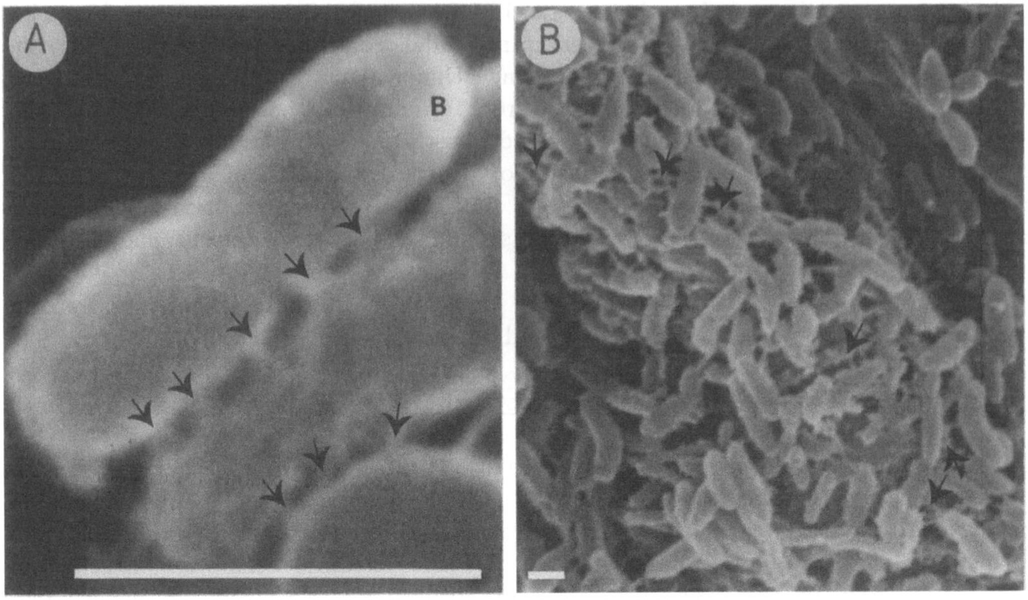

Fig. 2: Scanning electron microscopy of fibrillar material connections (A) between A. brasilense Cd cells, and (B) to the root surfaces. (Bars represent 1 μ)

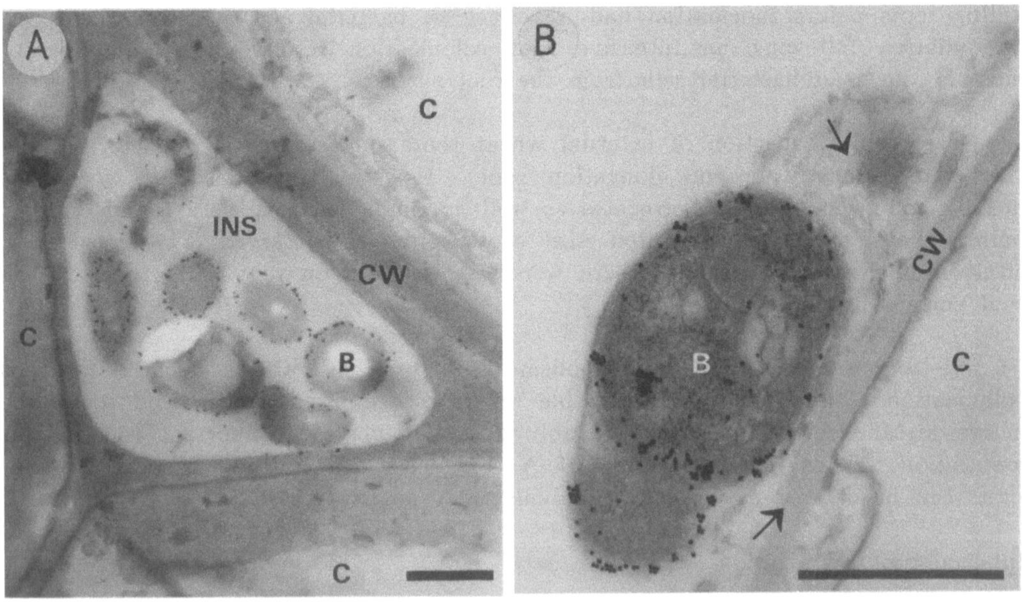

Fig. 3: (A) Transmission electron micrographs of specific identification of A. brasilense Cd cells within the intercellular spaces by coloidal-gold labeling.
(B) Binding of A. brasilense Cd cells to cortex intercellular spaces by an electıon dense material (arrows). (Bars represent 1 μ). Abbreviations: B - bacteria; C - cortex cells; CW - plant cell wall; INS - inter cellular spaces.

When an external bacterial population was established on the root surfaces an internal bacterial population developed too. Washings and strong agitation did not remove the internal population. Moreover, counts made by the specific enzyme-linked immunosorbent assay (ELISA), especially developed for the detection and enumeration of this strain (10), revealed that the internal bacterial population comprised the majority of the bacteria that colonized the roots. Cd population present on the roots, the internal bacteria comprised the majority of that population.

The population of A. brasilense Cd within wheat roots was detected by light and transmission electron microscopy, using the coloidal-gold immunocytochemical technique, which is based on specific recognition by antibodies (10) and labeling the cells with gold particles as described by Roth (22). The external and the internal root population of A. brasilense Cd was as intensively gold-labeled as its liquid culture, with a very low background. Other bacterial species on root surfaces or in intercellular spaces were not labeled at all. Cells of A. brasilense Cd were distributed along root surfaces, while the internal population was restricted to the elongation and to the root-hair zones. A. brasilense Cd was located in the cortex intercellular spaces, as deep as the endodermis, but did not penetrate either the endodermis or the vascular system (Fig. 3a). These tissues in corn plants were reported to be colonized by this genus (20). The bacteria were bound to cell walls in the cortex intercellular spaces as well as to epidermal cells by an electron dense material (Fig. 3b). Several cells of A. brasilense Cd were found within young, live root cells without any apparent damage to the plant cells.

We suggest that A. brasilense Cd forms an unharmful inter- and intra-cellular association with wheat roots, in addition to its root surface colonization.

Other effects of A. brasilense Cd inoculation on wheat roots

Several mechanisms and phenomena have been proposed to characterize inoculation by Azospirillum i.e., hormonal activity, nitrogen fixation, enhancement and improvement in mineral and water uptake, promotion of root hair development and root branching, alterations of cell arrangement of the cortex, increase in dry matter and in mineral accumulation in leaves and improvement of yields of cereals and forage grasses (1,2,5,17).

We present here other effects of A. brasilense Cd on wheat roots: (i) increase in root membrane excitation causing an increase in proton extrusion from the root to the rhizosphere (ii) enhancement of cell division in root tips.

We identified a physiological mechanism of proton pumping in wheat roots. The effect was measured as a decrease in pH at the root surrounding, using a pH

Fig. 4: Acidification of agar medium by wheat roots inoculated with A. brasilense Cd. Dark color represents pH 6.8; white color represents pH 3.7-4.5.

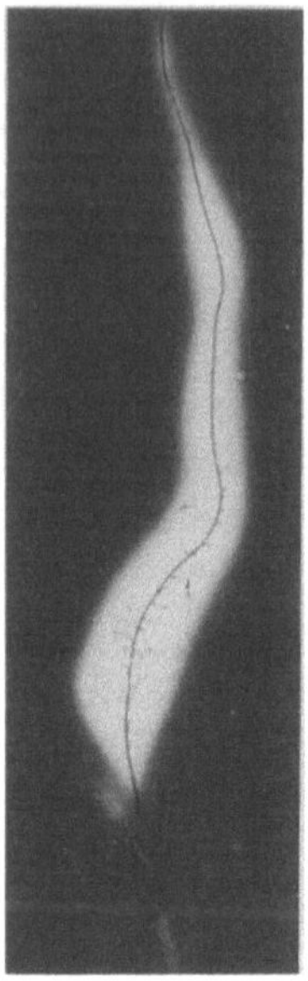

electrode, and visualized by the agar-indicator technique (23). Inoculation with A. brasilense Cd caused a measurable increase in the proton leakage from wheat roots (Fig. 4). Therefore, we concluded tht inoculation affects cell membranes, even in the absence of attachment of bacteria to plant membranes. This effect might be related to the known changes in mineral uptake in the plant after inoculation (24).

We determined the mitotic index (no. of cells at mitosis/total number of cells) of inoculated root tips (25), 72h after germination and inoculation. Analyses were carried out on several thousands cells in different roots. Significant increase in the mitotic index was observed in roots inoculated at 10^6 colony-forming unit/ml, which is known to be an optimal level of inoculation (6).

Conclusions

The biophysical and ecological features of A. brasilense demonstrated here, combined with the considerable progress made world-wide during the past decade, are not sufficient to account for the inconsistency in yield response obtained by many research groups. Further studies on these aspects are essential before inoculation technology can be released for commercial use.

Acknowledgements

This paper was written in memory of the late Mr. Avner Bashan. We thank Dr. Eugenia Klein and Mrs. Batia Romano from the Department of Biological Services, The Weizmann Institute of Science for helping in electron microscopy, Miss Dori Filon, Miss Einat Carmi, Mrs. Adi Ben-Yehuda and Mr. G. Richter-Levin for excellent technical assistance and Mr. Y. Avivi for editing the manuscript.

References

1. Gaskins, M.H., Albrecht, S.L. and Hubbell, D.H. (1985). Agric. Ecosystem Environ. 12,99-116.
2. Patriquin, D.G., Döbereiner, J. and Jain, D.K. (1983). Can. J. Microbiol. 20,900-915.
3. Schank, S.C. and Smith, R.L. (1984). Proc. Soil Crop Sci. Soc. Fla. 43,120-123.
4. Bashan, Y. (1986). Appl. Environ. Microbiol. 51,1067-1071.
5. Albrecht, S.L., Gaskins, M.H., Milam, J.R., Schank, S.C. and Smith, R.L. (1983). Experientia Suppl. 48,138-148.
6. Bashan, Y. (1986). Soil Biol. Biochem. 18,297-301.
7. Kapulnik, Y., Sarig, S., Nur, I., Okon, Y., Kigel, J. and Henis, Y. (1981). Exp. Agric. 17,179-187.
8. Smith, R.L., Schank, S.C., Milam, J.R. and Baltensperger, A.A. (1984). Appl. Environ. Microbiol. 47,1331-1336.
9. Bashan, Y. and Levanony, H. (1985). Can. J. Microbiol. 31,947-952.
10. Levanony, H., Bashan, Y. and Kahana, Z.E. (1987). Appl. Environ. Microbiol. 53,358-364.
11. Barak, R., Nur, I. and Okon, Y. (1983). J. Appl. Bact. 53,399-403.
12. Reinhold, B., Hurek, T. and Fendrik, I. (1985). J. Bact. 162,190-195.
13. Bashan, Y. (1986). J. Gen. Microbiol. 132,3407-3414.
14. Bashan, Y., and Wolowelsky, J. (1987). Soil Sci. 143,132-138.
15. Bashan, Y. (1986). Appl. Environ. Microbiol. 51,1089-1098.
16. Gafni, R., Okon, Y., Kapulnik, Y. and Fischer, M. (1986). Soil Biol. Biochem. 18,69-76.
17. Jain, D.K. and Patriquin, D.G. (1984). Appl. Environ. Microbiol. 48,1208-1213.
18. Umali-Garcia, M., Hubbell, D.H., Gaskins, M.K. and Dazzo, F.B. (1980). Appl. Environ. Microbiol. 39,219-226.
19. Schank,S.C., Smith, R.L., Weiser, G.C., Zuberer, D.A., Bouton, J.H., Quesenberry, K.H., Tyler, M.E., Milam, J.R. and Littell, R.C. (1979). Soil Biol. Biochem. 11,287-295.
20. Bashan, Y., Levanony, H. and Klein, E. (1986). J. Gen. Microbiol. 132,3069-3073.
21. Patriquin, D.G. and Döbereiner, J. (1978). Can. J. Microbiol. 24,734-742.
22. Matthews, S.W., Schank, S.C., Aldrich, H.C. and Smith, R.L. (1983). Soil Biol. Biochem. 15,697-703.
23. Roth, J. (1983). J. Histochem. Cytochem. 31,987-999.
24. Marschner, H., Römheld, V., Ossenberg-Neuhaus, H. (1982). Z. Pflanzenphysiol. Bd. 105,407-416.
25. Lin, W., Okon, Y. and Hardy, R.W.F. (1983). Appl. Environ. Microbiol. 45,1775-1779.
26. Darlington, C.D. and La Cour, L.F. (1962). The handling of chromosomes. George Allen and Unwin, London.

QUANTITATIVE MEASUREMENT OF AZOSPIRILLUM PLANT CELL ATTACHMENT

M. Eyers, F. Waelkens, J. Vanderleyden and A.P. Van Gool
F.A. Janssens Memorial Laboratory for Genetics
Catholic University of Leuven, Willem de Croylaan 42, B-3030 Heverlee

Summary: A quantitative plant assay system is described for measuring the attachment of *Azospirillum spp.* to plant cells. The attachment of different *Azospirillum* strains was assayed by mixing ^{14}C-leucine labeled bacteria with suspensions of freshly isolated *Zinnia elegans* leaf mesophyll cells. In these conditions *Azospirillum brasilense* Sp7 and Sp245 cells were found to attach to plant cells to an extent of 17 to 20 %. Heat killed *Azospirillum spp.* showed defective attachment.

Keywords: *Azospirillum*, attachment, *Zinnia elegans*, cell suspensions.

Introduction

It has been shown that bacteria of the genus *Azospirillum* can attach to root-hairs (1, 2, 3, 4) and that some strains even colonize the root interior of wheat and sorghum (4, 5).

A quantitative plant assay system to measure this attachment is necessary in order to reveal strain differences, to detect potential genetic determinants of *Azospirillum* involved in the plant root association and, eventually, to evaluate competition for plant root colonization by different rhizosphere bacteria. Such a plant assay system has been developed for other rhizosphere bacteria (*Agrobacterium*, *Rhizobium*). By means of genetic analysis of *Agrobacterium* Tn*5* mutants affected in attachment, determinants were characterized that are essential in plant root and plant cell attachment (6, 7, 8). The presence of some of these determinants of *Agrobacterium* and *Rhizobium* have been identified in *Azospirillum* by heterologous DNA:DNA hybridizations (9, 10). This could suggest a universal system for one or more steps involved in attachment. Using at random Tn*5* mutagenesis (11, 12) and site-directed Tn*5* mutagenesis (13), it will be possible to construct *Azospirillum* strains impaired in plant cell attachment and/or recognition.

Azospirillum IV: Genetics, Physiology, Ecology
Edited by W. Klingmüller

Materials and methods

Bacterial strains

The bacterial strains used in this study are listed in Table I.

Table I. Bacterial strains used

Strain	Relevant properties	Reference and/or source
E. coli HB101	*hsdS20* (r^-_B, m^-_B), *rec* A13	Maniatis *et al.*
Agrobacterium		
A. tumefaciens A208		Chilton *et al.*
A. tumefaciens A1011	Km^r *chv*::Tn*5*	Douglas *et al.*
Azospirillum		
A. brasilense Sp7	Ap^r	ATCC 29145
A. brasilense SP245	Ap^r	Döbereiner J.

Media and chemicals

Azospirillum spp. strains were grown at 28°C in yeast extract-peptone broth (YEP). *E. coli* HB101 was grown in L-broth at 37°C.

Plant cells

Zinnia elegans Jacq was grown from seed (Nationale Plantentuin, Meise). Seeds were germinated, and seedlings were grown in vermiculite. Leaves were rinsed in distilled water and then homogenized in Murashige-Skoog salts with a mortar and pestle. The homogenate was passed through a 250 μm screen (Bellco Glass, Inc.) and the cells were caught on a Miracloth filter (Calbiochem.). Cells were suspended in Murashige-Skoog salts to a concentration of 5×10^5 to 6×10^5 cells/ml, determined by counting with a hemacytometer. Suspensions obtained in this manner consisted mostly of single cells.

Assay of bacterial attachment to plant cells (Figure II)

Radio-active labeled bacteria were added to *Zinnia* leaf mesophyll cells and their attachment was monitored. *Zinnia* leaf mesophyll cells were prepared as described above immediately before use. Bacteria were labeled by inoculating a single colony into 0,5 ml of YEP medium supplemented with 2 uCi of L-[U-^{14}C]-leucine (342 mCi/mmol; Amersham International plc) and growing the culture overnight. The bacteria were washed twice in 10 mM $MgSO_4$. Specific activities of 10^{-5} cpm per cell were obtained. Labeled bacteria were added to plant cell suspensions at a final concentration of approximately 10^6 bacteria per ml, corresponding with a bacterium to plant cell ratio of approximately 3:1. The mixture was incubated at 28°C on a rotary shaker at 40 rpm for 2 h. The number of bacteria attached to plant cells was determined after filtering the bacteria-plant cell mixture through Miracloth filters. Free bacteria, but not plant cells with attached bacteria, passed through these filters. After a wash with 10 mM $MgSO_4$, the filters were placed in scintillation vials with 5 ml of scintillation liquid and counted in a liquid Tri-Carb scintillation counter.

Results

The results are summarized in Table II.

Live bacteria of *Azospirillum brasilense* Sp7 and Sp245 attach to *Zinnia* leaf mesophyll cells after 2 h of the addition of the bacteria to the plant cell suspensions. *Agrobacterium tumefaciens* A208 (wild type) and A1011 (mutant defective in plant cell attachment) were found to attach to the same extent as in the attachment experiments of C. Douglas (7). *E coli* HB101 was, as non-rhizosphere bacterium, used as control. Heat killed bacteria only showed a residual attachment (Table II). The binding of live bacteria and bacteria killed by exposure to heat (20 min at 65°C) was examined under the phase-contrast light microscope with 10^6 bacteria per ml and 10^5 *Zinnia* cells per ml (Figure I).

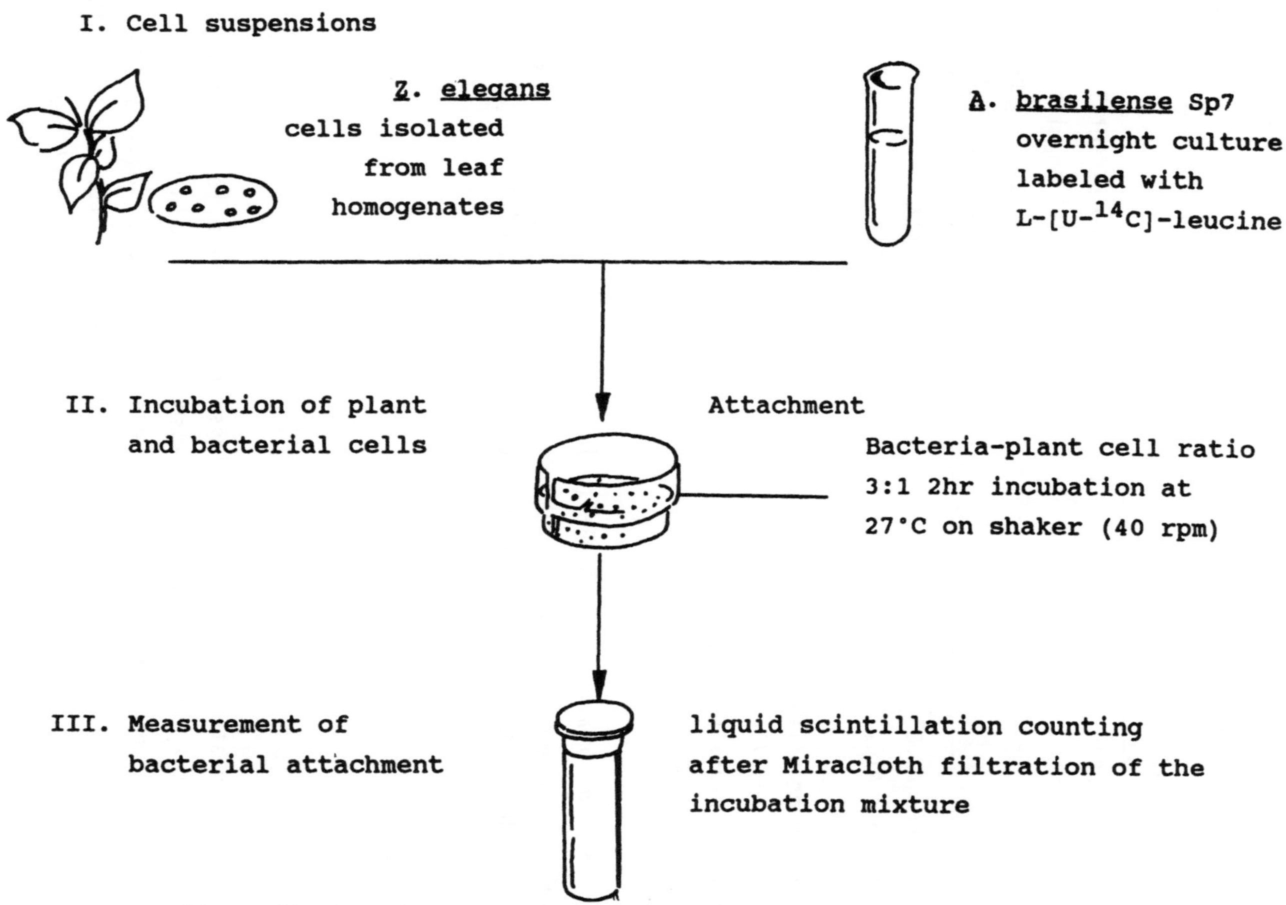

Figure II: Quantitative assay on the attachment of *Azospirillum* to *Zinnia elegans* mesophyll cells.

Table II: Attachment of *Azospirillum brasilense* strains to *Zinnia elegans* leaf mesophyll cells.

Bacterial strains	Fraction of total bacterial population attached to Z. elegans cells (replicates)
A. tumefaciens A208[1]	24/27/22/26
A. tumefaciens A1011[1]	5/4/7/8
E. coli HB101[1]	3/2/2/2
Azospirillum brasilense Sp7	19/20/17/19
Azospirillum brasilense Sp7 heat killed[2]	5
Azospirillum brasilense Sp245	17/18

[1] *Agrobacterium tumefaciens* A208 (wild type) and A1011, (chromosomal avirulent) mutant defective in plant cell attachment, (Douglas *et al.*, 1985, J. Bact. *152*, 1265-1275) were used as reference strains. *E. coli* HB101 was used as control.

[2] 20 min. at 60°C.

Discussion

We have demonstrated that *A. brasilense* cells attach to isolated plant cells of a dicot species, although *Azospirillum* is mainly found in the rhizosphere of monocots. *A. tumefaciens* cells attach equally well to isolated plant cells of monocots as dicots (14), despite the fact that monocots are no host plants for *A. tumefaciens*. This could suggest that bacteria-plant cell attachment, to a first approximation, is directed by the bacterial partner and that these bacterial determinants are not plant species specific.

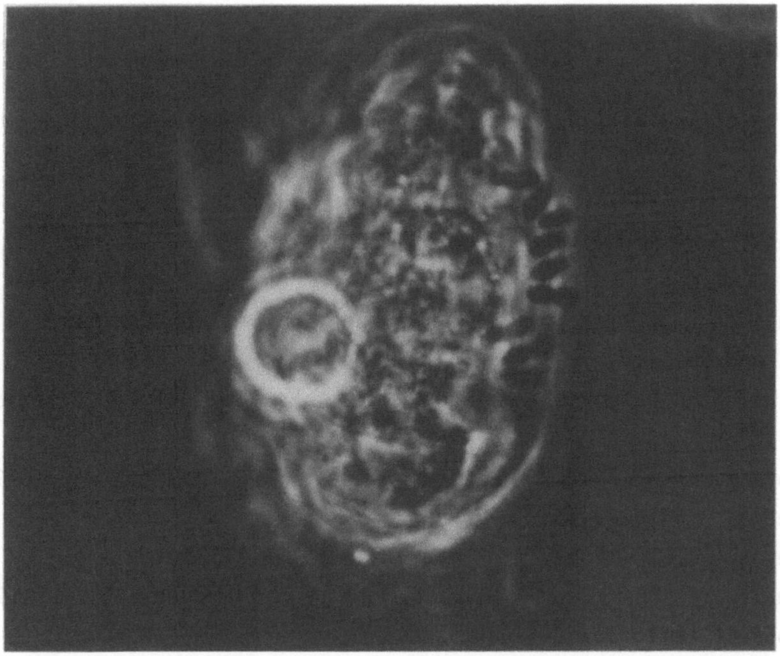

Figure I: Light micrograph of a *Zinnia* leaf mesophyll cell in the presence of *Azospirillum brasilense* Sp7.

The molecular mechanisms of bacteria-plant attachment are not understood, but at least from *Agrobacterium*, genetic loci required for attachment have been identified (6, 8) and cloned (6). DNA hybridizations between some of these loci and total DNA of *Azospirillum* have revealed homology (10). Experiments in progress to demonstrate that these sequences are indeed functional in *Azospirillum*, together with the hereby described results, make it possible to construct and to test *Azospirillum* strains, either impaired or improved for plant cell attachment.

Acknowledgments

We gratefully acknowledge J. Döbereiner and Drs. C. Douglas for providing strains. M.E. is the recipient of a fellowship from the IWONL. F.W. is the recipient of a fellowship from the NFWO.

References

1. Bashan, Y., Levanony, H. and Klein, E. (1986). J. of Gen. Microbiology *132*, 3069-3073
2. Jain, D.K. and Patriquin, D.G. (1984). Appl. Environm. Microb. *48*, 1208-1213
3. Kapulnik, Y. (1981). Plant Physiology *68*, 340-343
4. Baldani, V., Baldani, D., Ivo Baldani, J. and Döbereiner, J. (1983). Can. J. of Microbiology *29*, 924-929
5. Baldani, V., Alvarez, M., Ivo Baldani, J. and Döbereiner, J. (1986). Plant and Soil *90*, 35-36
6. Douglas, C., Halperin, W. and Nester, E. (1982). J. Bacteriol. *152*, 1265-1275
7. Puvanesarajah, V., Schell, F., Stacey, E., Douglas, C. (1985). J. Bacteriol. *164*, 102-106
8. Matthysse, A.G. (1986). J. Bacteriol. *169*, 313-323
9. Fogher, C., Dusha, I., Barbot, P. and Elmerich, C. (1985). FEMS Microbiol. Lett. *30*, 245-249
10. Waelkens, F., Maris, M., Verreth, C., Vanderleyden, J., Mergeay, M. and Van Gool, A.P. (1987). FEMS Microbiol. Lett., in press
11. Singh, M. and Klingmüller, W. (1986). Mol. Gen. Genet. *202*, 136-142
12. Vanstockem, M., Michiels, K., Vanderleyden, J. and Van Gool, A.P. (1987). Appl. Environm. Microbiol. *53*, 410-415
13. Perroud, B., Bandhari, S.K. and Elmerich, C. (1985). In Klingmüller, W. (ed.) *Azospirillum* III: Genetics, Physiology, Ecology. Springer-Verlag, Heidelberg, p. 10-19
14. Douglas, C., Halperin, W., Gordon, M. and Nester, E. (1985). J. of Bacteriol. *161*, 764-766
15. Neyra, C. (1977). Adv. Agron. *29*, 1-38

EFFECTS OF INOCULATION WITH DIFFERENT STRAINS OF AZOSPIRILLUM BRASILENSE ON WHEAT ROOTS DEVELOPMENT

P. Barbieri*, A. Bernardi*, E. Galli* and G. Zanetti**
*Dipartimento di Genetica e di Biologia dei Microrganismi
**Dipartimento di Fisiologia e Biochimica generali
Università degli Studi di Milano, via Celoria 26 - 20133 Milano - Italy

Summary: Inoculation of wheat seedlings (Triticum durum var. Appula) with different wild type strains of Azospirillum brasilense significantly increased the number and length of lateral roots, even if the increase due to the inoculation with strain Sp6 was always higher than that obtained with strains Sp7 and F14; instead, inoculation with strain 7310, a derivative of Sp7, unable to produce indole-3-acetic acid, did not elicit any response from the plant. Control uninoculated experiments in which different concentrations of indole-3-acetic acid were added, apparently showed a dose-effect on root system development. Root system was affected by the inoculum concentration: 10^6 cfu/ml caused the greatest increase in the number and length of lateral roots, while 10^9 cfu/ml inhibited root development. Keywords: indole-3-acetic acid, wheat root system development.

Introduction

It is known that seeds inoculation with Azospirillum strains produces beneficial effects on crop plants (1,2). Inoculation of wheat seedlings with Azospirillum brasilense Sp6 or Azospirillum brasilense SpF103, a Nif^- mutant of Sp6, both producers of indole-3-acetic acid, caused an increase in the number and length of lateral roots, whereas inoculation with SpF57, a Nif^- strain low producer of indole-3-acetic acid, did not affect the root

Azospirillum IV: Genetics, Physiology, Ecology
Edited by W. Klingmüller

development (3). These data suggested that the root system response was due mainly to the production of phytohormone-like substances by Azospirillum strains.

In the present work we describe further inoculation experiments of wheat seedlings with different strains of A. brasilense and report the influence of inoculum concentration on the root system development. Preliminary results obtained from experiments in which indole-3-acetic acid was directly supplied in the absence of inoculation, are also presented.

Materials and Methods

Bacterial strains and cultural conditions. Three wild type strains of Azospirillum brasilense, Sp6 (4), F14 and Sp7 (ATCC 29145), and a derivative of strain Sp7, 7310, were used in the present work. Strain F14, isolated by wheat roots, was kindly supplied by F. Favilli, and strain 7310, which shows a reduced nitrogenase activity, by C. Elmerich. Minimal medium MSP and growth conditions were as described (5). Isolation and determination of indole-3-acetic acid (IAA) by HPLC were performed as previously reported (3).

Inoculation experiments. Inoculation experiments on wheat seeds (Triticum durum var. Appula) were performed as previously described (3). Bacterial inocula were grown in NFb medium (3), washed twice in phosphate buffer 0.05M, pH 7, resuspended in sterile distilled water and inoculated in the test tubes to give the required final concentration (10^4 to 10^9 cfu/ml). IAA for inoculation experiments was dissolved in methanol, appropriately diluted in distilled water, filtered through a 0.45 µm membrane filter Millipore Millex-HA and added to the test tubes to give a final concentration of 10^{-10}, 10^{-11}, 10^{-12} M respectively. Tubes were incubated at 25°C in an illuminated room for 14 days. At the end of the incubation period, the number and length of lateral roots, the length of main roots, the fresh weight of roots and shoots were measured and the bacterial population, free in solution and associated to the roots, was determined by plate count technique on MSP medium.

Results and Discussion

Indole-3-acetic acid produced by Azospirillum strains Sp6, F14, Sp7 and 7310 was measured in cultural broth, using tryptophan as precursor of IAA (6). As shown in Table 1, F14 and Sp7 produced a greater amount of IAA than Sp6, whereas no IAA was detectable in the culture broth of 7310.

TABLE 1. Indole-3-acetic acid production by A. brasilense strains.

Strain	Growth conditions[a]	IAA ($\mu g/A_{670}$/7 days)
Sp6	N-free medium[b]	9
Sp7	N-free medium[b]	24
F14	N-free medium[b]	24
7310	NH_4Cl 2.5mM[b]	nd

[a] The cultures were supplied with tryptophan (100 µg/ml)
[b] Static cultures
nd: not detectable

Inoculation experiments were carried out in a hydroponic closed system without addition of tryptophan. After 14 days of incubation the parameters reported in Table 2, 3, 4 were measured. As shown in Table 2, 3, we compared the effects on the root system development of inoculation with Sp6, Sp7 or F14. Inoculum concentration of bacteria in the test tubes was 10^6 cfu/ml. In all cases inoculation produced a significant increase in the number and length of lateral roots, however the effects produced by Sp6 were significantly higher than those induced by strains Sp7 and F14. At the end of the experiments counts of bacteria free in solution and associated to the roots were respectively 10^6 cfu/ml and 10^7 cfu/g. When inoculation experiments were performed with strain 7310 (Table 4) no significant differences were found in the growth of the root system between controls and inoculated samples. Such findings are in agreement with data previously obtained with strain SpF57 (3).

Preliminary experiments performed by adding different concentrations of IAA to the uninoculated test tubes suggested a dose-effect on the root system development, i.e. number and length of lateral roots and length of main root; the optimal concentration seemed to be in the range 10^{-11}-10^{-12} M.

To investigate the effects of inoculum concentration, as suggested by Kapulnik et al. (7), experiments with strains Sp6 and Sp7 were carried out. The concentration that caused the greatest increase in the number and length of lateral roots was 10^6 cfu/ml, while 10^9 cfu/ml caused inhibition of root development (Fig. 1). No effect was observed when the samples were inoculated with 10^4 cfu/ml.

At the end of the experiments bacteria associated to the roots were 10^8 cfu/g for the lowest inoculum concentration and 10^9 cfu/g for the highest. Similar results were obtained with strain F14 (data not shown). Such data indicate that plant roots respond differently to different levels of Azospirillum inoculation, suggesting an effect on the root colonization, indipendently from the different levels of IAA produced by the tested strains (Table 1). Our findings agree with the results reported by Kapulnik et al. (7), although they showed that a lower inoculum concentration was required when inoculation was performed with an IAA over-producing mutant which produced about 30 times more IAA than the wild type. In our experiments we have always found that the optimal inoculum concentration was 10^6 cfu/ml, but strains Sp7 and F14 produced only about 3 times more of IAA in respect to Sp6.

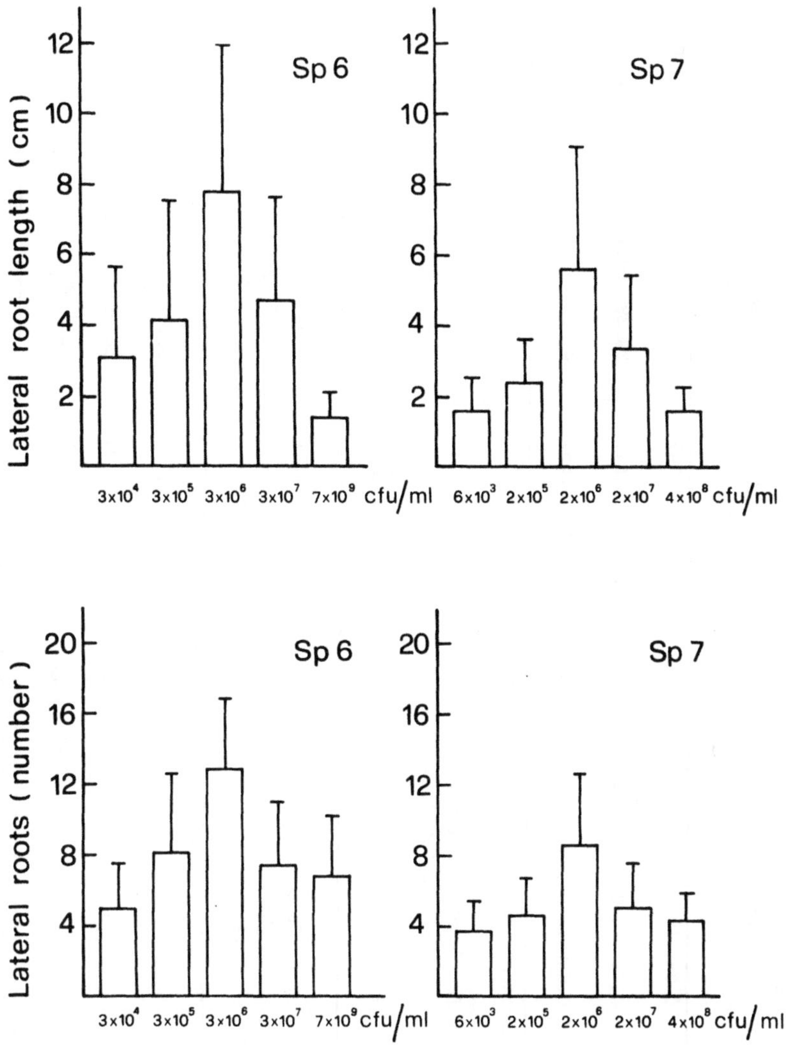

Fig. 1: Effect of different inoculum concentrations on root system development of Triticum durum var. Appula. Data represent mean of 18 replicates for inoculation with Sp6 and of 16 replicates for inoculation with Sp7.

TABLE 2. Effect of inoculation with of A. brasilense Sp6 and Sp7 on the root system of Triticum durum var. Appula.

Conditions	Strain	Root fresh weight (g)	Lateral roots (number)	Lateral root length (cm)	Main root length (cm)	Shoot fresh weight (g)
Uninoculated	-	0.041 (0.01)	5.54 (5.09)	3.03 (2.04)	40.82 (9.93)	0.143 (0.03)
Inoculated	Sp6	0.045 (0.01)	16.20[a] (8.41)	14.47[a] (8.84)	39.39 (9.20)	0.149 (0.03)
Inoculated	Sp7	0.045 (0.01)	10.88[ab] (8.48)	6.09[ab] (6.65)	42.78 (11.60)	0.153 (0.03)

Data represent means of 24 replicates. Standard deviation is given in parentheses.

[a] Values significantly different from control ($P \leq 0.05$)
[b] Values significantly different from Sp6 ($P \leq 0.05$)

TABLE 3. Effect of inoculation with A. brasilense Sp6 and F14 on the root system of Triticum durum var. Appula.

Conditions	Strain	Root fresh weight (g)	Lateral roots (number)	Lateral root length (cm)	Main root length (cm)	Shoot fresh weight (g)
Uninoculated	-	0.048 (0.01)	9.00 (4.22)	3.33 (1.79)	35.67 (5.44)	0.143 (0.03)
Inoculated	Sp6	0.047 (0.01)	21.48[a] (9.23)	17.93[a] (15.44)	50.98[a] (7.91)	0.138 (0.03)
Inoculated	F14	0.045 (0.01)	15.45[ab] (8.82)	9.7[ab] (6.06)	42.64[ab] (12.88)	0.137 (0.02)

Data represent means of 27 replicates. Standard deviation is given in parentheses.

[a] Values significantly different from control ($P \leq 0.05$)
[b] Values significantly different from Sp6 ($P \leq 0.05$)

TABLE 4. Effect of inoculation with <u>A. brasilense</u> 7310 on the root system of <u>Triticum durum</u> var. <u>Appula</u>.

Conditions	Strain	Root fresh weight (g)	Lateral roots (number)	Lateral root length (cm)	Main root length (cm)	Shoot fresh weight (g)
Uninoculated + NH_4Cl 2.5mM	-	0.040 (0.01)	3.08 (2.46)	0.30 (0.24)	25.25 (3.67)	0.126 (0.03)
Inoculated + NH_4Cl 2.5mM	7310	0.032 (0.01)	3.58 (2.78)	0.73 (0.57)	25.07 (2.78)	0.123 (0.03)

Data represent means of 12 replicates. Standard deviation is given in parentheses.

Acknowledgements

We thank Dr. Favilli and Dr. Elmerich for kindly providing Azospirillum strains. This study was supported by the Consiglio Nazionale delle Ricerche, Italy, special grant I.P.R.A. sub-project 1, paper n° 1406.

References

1) Döbereiner, J. (1983). In Klingmüller, W., (ed.) Azospirillum II: Genetics, Physiology, Ecology. Experentia Supplementum 48, Birkhäuser-Verlag, Basel, p. 9-23
2) Okon, Y. (1985). Trends Biotechnol. 3, 223-228
3) Barbieri, P., Zanelli, T., Galli, E. and Zanetti, G. (1986). FEMS Microbiology Letters 36, 87-90
4) Bani, D., Barberio, C., Bazzicalupo, M., Favilli, F., Gallori, E. and Polsinelli, M. (1980). J. Gen. Microbiol. 119, 239-244
5) Ratti, S., Curti, B., Zanetti, G. and Galli, E. (1985). J. Bacteriol. 163, 724-729
6) De Francesco, R., Zanetti, G., Barbieri, P. and Galli E. (1985). In Klingmüller, W., (ed.) Azospirillum III: Genetics, Physiology, Ecology. Springer-Verlag, Heidelberg, p. 109-115
7) Kapulnik, Y., Okon, Y. and Henis, Y. (1985). Can. J. Microbiol. 31, 881-887

NITROGENASE ACTIVITY OF STERILE CEREAL SEEDLINGS INOCULATED WITH AZOSPIRILLUM BRASILENSE: DIFFERENCES BETWEEN CULTIVARS AND POSSIBLE SIGNIFICANCE

G. Jagnow

Institut für Bodenbiologie, Bundesforschungsanstalt für Landwirtschaft, Bundesallee 50, D-3300 Braunschweig, FRG

Summary: Large differences in supporting nitrogenase activity existed between 14 German cereal cultivars. Fractions of active plants ranged between 3 and 67% and maximal activities between 5 and 35 nmol $C_2H_4.h^{-1}$. No significant differences were found in harvest parameters, root populations of *Azospirillum* and fractions of active F 1 plants between nitrogenase positive and negative wheat plants. Ear emergence of active plants was accelerated. Keywords: cereals, cultivars, *Azospirillum*, nitrogenase,

Introduction

Natural differences in nitrogenase (N_2-ase) activity of their rhizosphere were reported from cultivars of rice (14, 16), *Brachiaria* spp. (15) and sugar cane (19). Similar differences were also found between cereal cultivars using sterile, inoculated seedlings of rice (5), *Sorghum bicolor* (11), maize (6) and *Pennisetum americanum* (4, 20). In a similar seedling test, large differences were also found between three German wheat cultivars (9). N_2-ase activities of seedlings of additional German cereal cultivars, the inheritance of positive or negative "nis" characters (nitrogen fixation supporting), growth characteristics and *Azospirillum* root populations of nis-positive and negative plants will be reported here.

Materials and methods

Seeds from nine wheat-, four barley- and one rye cultivars were obtained from our department of agronomy and plant breeding.

Azospirillum IV: Genetics, Physiology, Ecology
Edited by W. Klingmüller

Seeds were sterilized, grown in N-free semisolid mineral agar and inoculated as described previously (1, 9, 18). For inoculations, A. brasilense (ATTC 29 710) or A. lipoferum isolated locally (7) were used, both resulting in the same ranges of activities. N_2-ase activities were determined by the acetylene reduction test (8).

Nineteen active (8-34 nmol $C_2H_4.h^{-1}$) plus 14 inactive seedlings of summer wheat "Max" and 29 active (6-20 nmol $C_2H_4.h^{-1}$) plus 24 inactive seedlings of summer barley cv. "Koral" were transplanted after testing into 13 cm plastic pots with a mixture of 50% v/v of a sieved parabrown earth (8) and 50% v/v of a porous granulate of perlite, grown in a greenhouse and irrigated with tap water. After 30 and 60 days a modified Hoagland nutrient solution was given (18), supplying an equivalent of 40, 9 and 11 $kg.ha^{-1}$ of N, P and K, respectively. After ear emergence, from each of the positive plants of summer barley a root line was sampled after removing root bales and the bales replanted into the pots again. Rhizosphere populations of Azospirillum were determined (8). Before flowering the ears of pairs of active wheat plants with similar seedling activities were enclosed in paper bags and grown to maturity. After harvest dry matter and N-contents were determined. Grains were harvested separately from each ear, sterilized, germinated, inoculated and assayed for N_2-ase activity as their parent plants.

Results

Seedlings of the 14 cereal cultivars showed large differences both with respect to their fractions of active plants and to their maximal N_2-ase activities, respectively (table 1). The eight summer wheat cultivars were most different with 3-68% active and 3-38% strongly active plants (5 nmol $C_2H_4.h^{-1}$). The summer barley cultivars "Golda", "Koral" and "Georgi" together with the summer wheat cultivars "Max", "Amor", "Turbo", "Kokat" and "Achill" were the most active. Fractions of active plants and activities of other cultivars were distinctly lower. Activities

Tab. 1. Frequency and rate of nitrogenase activity of sterile seedlings of German cereal cultivars after inoculation with *Azospirillum brasilense* or *A. lipoferum*

	number of plants	% of active plants >0.2nmol/h	>5nmol/h	Maximal N_2 ase-activity (nmol C_2H_4/h
Summer wheat:				
"Max"	63+	68_a ++	38_a	34.2
"Amor"	60	57_{ab}	23_{ab}	21.1
"Turbo"	60	50_{ab}	17_{bc}	12.4
"Kokat"	50	48_b	14_{bc}	12.0
"Achill"	12	42_{bc}	25_{ab}	42.2
"Horizont"	60	37_{bc}	5_{bc}	13.3
"Sokrates"	47	36_{bc}	6_{bc}	11.0
"Kadett"	30+	3_d	3_{bc}	7.3
Winter wheat:				
"Disponent"	51+	25_c	6_{bc}	13.5
Summer barley:				
"Golda"	62	63_{ab}	27_{ab}	33.6
"Koral"	55	60_{ab}	16_{bc}	20.8
"Georgi"	62	56_{ab}	26_{ab}	35.4
Winter barley:				
"Diana"	67	46_b	10_b	10.6
Summer rye:				
"Sorom"	62	24_{bc}	3_c	5.6

\+ inoculated with *A. lipoferum*

++ values with different index letters are significantly different (P = 0.05 or less) according the chi-square test.

within cultivars generally seemed to follow log-normal distributions (9).

During pot cultivation active and inactive wheat and barley seedlings showed no distinct differences. Ear emergence of active plants took place two days earlier. There was no correlation between the N_2-ase activity of barley seedlings and their **Azospirillum** root populations after 75 days of pot cultivation. Active plants of both wheat and barley had lower shoot dry weights and N contents, but higher root dry weights (7 and 16% with wheat and barley, respectively) and higher 1000 grain-weights (7 and 10%, respectively).

Fractions of N_2-ase positive F 1-seedlings from enclosed ears of parent plants could be very different (table 2). The sets of F 1-seedlings from ears of both active and inactive parent plants could contain no, few or larger numbers of active plants.Similar large differences could be found among F 1-seedlings from ears of inactive plants enclosed together. The pooled fractions of active F 1-plants from active and inactive parent plants were 15.5 and 17.4%, respectively, and statistically not significant.

Discussion

Variability and genetics of the nis-character was investigated so far only with maize (6) and Pennisetum spp. (3, 4, 13). First results about German wheat and barley cultivars are presented here. The lines of Pennisetum americanum had different levels of N_2-ase activity, but no negative plants (3, 4), whereas our wheat and barley cultivars had 32-97% and 37-54% of negative plants, respectively. With lines of P. americanum different, distinct levels of N_2-ase activity were found between 3 and 23 nmol $C_2H_4.h^{-1}$, but German cultivars had a rather continuous spectrum of nis-positive plants with activities log-normally distributed between 0.2 and 42 nmol $C_2H_4.h^{-1}$ and maximal activities distincty higher than the most active lines of P. americanum. Since our cultivars represented no distinct lines and were never selected for the nis-character, they can not be exspected to be uniform in this

Tab. 2. Frequency of nis-active plants among seedlings from parent plants with different nitrogenase activity as sterile seedlings of summer wheat cv. "Max" after inoculation with A. lipoferum

parent activity (nmol/h)	F1 plants per ear	% active F1 plants	parent activity (nmol/h)	F1 plants per ear	% active F1 plants
34.2	30[1)]	$23^{2)}_{ab}$	0	24[1)]	$4^{2)}_{a}$
32.3	31	22_{ab}	0	24	4_{a}
34.6	30	30_{b}	0	24	12_{ab}
30.0	33	18_{a}	0	24	4_{a}
25.4	32	0_{a}	0	24	29_{ab}
21.0	27	0_{a}	0	24	0_{a}
17.3	36	3_{ab}	0	24	42_{bc}
16.1	34	6_{ab}	0	24	8_{ab}
14.2	17	0_{a}	0	24	0_{a}
13.2	23	0_{a}	0	24	0_{a}
13.2	29	0_{a}	0	25	40_{bc}
12.4	24	0_{a}	0	33	39_{c}
10.9	31	23_{ab}	0	39	31_{bc}
10.5	34	3_{ab}	0	42	14_{ab}
10.4	31	26_{b}			
7.3	41	63_{c}			
8.1	40	18_{ab}			

1) rows with closer spacing contain results from ears enclosed together in bags for cross-pollination.

2) percentage values with different index letters are significantly different with P = 0.05 or less according to the chisqare test.

trait. Nevertheless, different fractions of active plants and ranges of activities seem to be characteristic for different cultivars.

A feature of nis-positive plants was earlier heading, which was also observed with more strongly nis-positive lines of P. typhoides (13). This may be a characteristic of plant genotype linked with the nis character, but it may be also a consequence of young active plants being colonized with Azospirillum more densely. They may have been more exposed to microbial growth regulatory substances than nis-negative plants. Accelerated heading, increases in tillering, root growth and branching and yield were reported after inoculation of summer wheat with A. brasilense and thought to be caused mainly by microbially formed plant growth regulators (2, 10). Indole acetic acid (IAA) was found in cultures of an over-producing A. brasilense, and IAA sprayed on wheat roots in soil at the same concentration had the same effect on root elongation as inoculation with this strain (12). No correlation was found between seedling N_2-ase activity of barley and the Azospirillum root populations. In Israel, wheat roots inoculated with A. brasilense Cd in unsterile soil showed high populations of this organism only initially. But they dropped by 90% or more within two weeks as a consequence of competing microorganisms unless these were checked by microbicidal substances not inhibitory to A. brasilense Cd (2). It is possible that a correlation between N_2-ase activity and Azospirillum populations might have existed in the rhizosphere of monoxenic plants, but was obscured by the highly variable competing soil microorganisms.

Differences in nis-positive fractions between F 1 plants from ears enclosed together suggest self-pollination rather than true crossing (table 2). Furthermore, descendants from both active and inactive parent plants had low and high proportions of nis-positives. Their parent plants did not represent pure lines, and since the F 1 generation was only a first step to genetic segregation one should exspect still a rather heterogenous outcome with respect to "nis". It seems likely, that the nis feature is determined by at least two, if not more genes. Breeding analysis of

P..typhoides suggested that both dominant and recessive genes were responsible for nis-expression (13). If with our cereals rhizosphere manipulation by introduction of growth-promoting bacteria and/or VA-mycorrhiza fungi should become economically feasable (21), their ability to support beneficial rhizosphere bacteria by their root exudates, as exemplified by the "nis" feature, should also have been analyzed genetically and optimized in breeding programs.

Acknowledgements:

This work was supported by the Deutsche Forschungsgemeinschaft. The author thanks Mrs. Susanne Scheil and Mrs. Ilse Schmidt for their skilled assistance.

References

1. Barber, D.A. (1967). J. Exp. Bot. 54, 163-169
2. Bashan, Y. (1986). Appl. Envir. Microbiol. 51, 1067-1071
3. Bouton, J.H. and Brooks, C.O. (1982). Crop Sci. 22,680-682
4. Bouton, J.H., Albrecht, S.L. and Zuberer, D.A. (1985). Field Crops Res. 11, 131-140
5. Charyulu, P.B.B.N., Fourcassie, F., Barbouche, A.K., Rondro Hariosa, L., Omar, A.M.M., Weinhard, P., Marie, R. and Balandreau, J. (1985). In Klingmüller, W. (ed.) Azospirillum III: Genetics, Physiology, Ecology. Springer-Verlag, Heidelberg, p. 163-179
6. Ela, St.W., Anderson, M.A. and Brill, W.J. (1982). Plant Physiol. 70, 1564-1568
7. Jagnow, G. (1982). In Klingmüller, W. (ed.) Azospirillum: Genetics, Physiology, Ecology. Experientia Supplementum 42, Birkhäuser-Verlag, Basel, p. 100-107
8. Jagnow, G. (1983). Z.Pflanzenernähr. Bodenk. 146, 217-227

9. Jagnow, G. (1985). In Klingmüller, W. (ed.) Azospirillum III: Genetics, Physiology, Ecology. Springer-Verlag, Heidelberg, p. 203-214

10. Kapulnik, Y., Feldman, M., Okon, Y. and Henis (1985). Soil Biol. Biochem. 17, 509-515

11. Kiepe-Nolt, J.A., Avalakki, U.K. and Dart, P.J. (1985). Soil Biol. Biochem. 17, 859-863

12. Kolb, W. and Martin, P. (1985). In Klingmüller, W. (ed.) Azospirillum III: Genetics, Physiology, Ecology. Springer-Verlag, Heidelberg, p. 215-221

13. Manga, V.K., Venkateswarlu, B. and Saxena, M.B.L. (1985). Ind. J. Agric. Sci. 55, 391-392

14. Ohta, K, Sano, Y., Fuji, T. and Iyama Shin-ya (1984). Japan. J. Breed. 34, 29-34

15. Pereira, P.A., Döbereiner, J. and Neyra, C.A. (1981). Can. J. Botany 59, 1475-1479

16. Rao, J.L.N. and Rao, V.R. (1985). Current Sci. 54, 752-754

17. Rennie, R.J., de Freitas, J.R., Ruschel, A.P. and Vose, P.V. (1983). Can. J. Bot. 61, 1667-1671

18. Robbins, W.J. and Kavanagh, F. (1938). Amer. J. Bot. 25, 229-236

19. Ruschel, R. and Ruschel, A.P. (1981). In Vose, P.B. and Ruschel, A.P. (eds.) Associative Nitrogen fixation, Vol. II, CRC Press, Cleveland, p. 133-141

20. Venkateswarlu, B. and Rao, A.V. (1985). Proc. Ind. Acad. Sci. (Plant Sci.) 95, 237-245

21. Whipps, J.M. and Lynch, J.M. (1986). In K.C. Marshall (ed.) Advances in Microbial Ecology 9, 187-244

IMPORTANCE OF NONSYMBIOTIC NITROGEN-FIXING BACTERIA IN ORGANIC FARMING SYSTEMS

P. Markus and J. Krämer

Institut für Pflanzenkrankheiten, Universität Bonn

Abteilung Landw. und Lebensmittel-Mikrobiologie

Meckenheimer Allee 168, 5300 Bonn

Summary: In organic cultivated soils common nitrogen-fixing bacteria like *Azotobacter* appeared in higher densities than *Azospirillum*. Establishment of inoculated *Azospirillum* strains S1 and S12A were successful with wheat. The use of organic manure A supported establishment of *A. lipoferum* S1. To investigate the role of cellulytic microorganisms (e.g. *Cellulomonas uda*) in enhancing nitrogen-fixing *A. lipoferum* S1, they were grown in mixed and pure cultures on cellulose and organic manure A. Growth and nitrogenase activity were stimulated in mixed cultures, different with cellulose and organic manure A. Keywords: Azospirillum, Cellulomonas, organic substances, wheat.

Introduction

A major characteristic of organic farming systems is the use of organic fertilizers instead of mineral nitrogen fertilizers. This additional carbon source could support nitrogen-fixing activity of soil microorganisms because of the limitation of carbon for nitrogen fixation in greater amounts (Postgate, 1982).

The goals of our studies was to determine whether:

1. bacterial density and activity can be stimulated by organic fertilizers,
2. there is an interaction between inoculated *Azospirillum spp.* and other rhizosphere microorganisms, especially cellulytic microorganisms.

Azospirillum IV: Genetics, Physiology, Ecology
Edited by W. Klingmüller

Material and Methods

Nitrogen-fixing (acetylene reduction) activity:
Acetylene-reduction activity of wheat-soil cores (15x3,5 cm) were determined immediately after sampling. Samples were placed in plastic bags, sealed, evaluated and refilled with 10% acetylene and air. Ethylene production was measured gaschromatographically after 24 h incubation at 25°C and corrected for soilborne ethylene after Norstaedt (1984).

Pot experiments:
Experiments were conducted to determine the influence of nitrogen-fixing *Azospirillum* strains (S1, S12A) in combination with different organic fertilizers on wheat plant growth and yield.
Soil characteristics: The soil was collected from organic cultivated soil and had the following makeup: sandy loam, pH-value 6.2 (0.01 m $CaCl_2$), total carbon content 1.32%, total nitrogen content 0.01%, 31 mg P_2O_5 (DL) 100 g soil and 18 mg K_2O (DL)/100 g soil.
Strains: The nitrogen-fixing strains of *Azospirillum spp.* (S1 and S12A) were isolated from the rhizosphere of wheat and characterized according to Tarrand et al. (1978).
Organic Manure : The mineral analysis of the two organic manures A and B were: total carbon content 12.3% (A) and 29.0% (B), total nitrogen content: 0.90% (A) and 1.95% (B), C_t:N_t ratios: 13.7% (A) and 14.9% (B).
This trial was run as a two-factorial experiment in four replicates over one vegetation period with normal light intensity and temperatures under a glass top. Six plants were grown per pot (3.3 kg).

Liquid culture systems:
Experiments were conducted in semisolid nitrogen deficient medium (mod. after Baldani and Döbereiner, 1979; Halsall and Goodchild, 1986) for 18 days to evaluate the role of cellulytic bacteria in simultaneous culture with nitrogen-fixing *Azospirillum sp.* and different organic sources.
Strains: Nitrogen-fixing *Azospirillum lipoferum* S1, isolated from the rhizosphere of wheat and *Cellulomonas uda*, (ATCC 491) isolated from compost, were examined.

Organic sources: Organic manure (A) and Cellulose (Sigmacell, type 20) 10 g/l with a small amount of $(NH_4)_2$ SO_4 (2 mg/l) to initiate bacterial growth.
These experiments were run in glass tubes (15 ml) containing 5 ml of the above medium.
Cultivar: In pot and liquid culture experiments the wheat cultivar Ralle was used.
Inoculation procedure: In pot experiments wheat seeds were surface-sterilized with Chlorox (1%, 10 min), then washed several times with sterile water and inoculated by dipping with bacteria suspension (10^{6}-10^{9} cells/ml suspension). Controls were treated with noninoculated medium.
Bacterial isolation in field and pot experiments: Bacteria were isolated from the root system of wheat plants. The root samples were macerated, suspended in Ringer solution (Oxoid) and diluted tenfold.
Bacterial counts: All counts of Azospirillum spp. were done with the MPN-method (Okon et al., 1977) in pot experiments and with a plate count method (TSA, Oxoid) in liquid culture experiments. Counts of Azotobacter spp. were done on selective agar after Gadkari and Stolp (1984).

Results and Discussion

In organic cultivated soils relatively low numbers of *Azospirillum* spp. could be found in the rhizosphere of wheat (Table 1). Common nitrogen-fixing bacteria, like *Azotobacter* spp., appeared in much higher densities. But nevertheless effective nitrogen-fixing *Azospirillum* spp. were isolated and used as effective inoculants in temperate climates (Jagnow, 1985).
The nitrogen-fixing (acetylene reduction) capacity under field conditions varied between 20.6 to 29 nmol evolved ethylene/cereal-soil core and was lower compared to nitrogen-fixing activity of clover-soil cores ($\bar{x}$ 3.74 umol evolved ethylene).

Beneficial effects were reported with the use of organic material in inoculation procedure (Ishac et al., 1982) and as organic fertilizer supporting nitrogenase activity in soil (Reents, 1987). The *Azospirillum* spp. numbers after inoculation with strain S1 and S12A in combination with two

Table 1: Densities of Azospirillum and *Azotobacter* spp. in the rhizosphere of wheat in organic cultivated soils

Range of *Azospirillum* spp./g root dry weight	Range in numbers of *Azotobacter* spp./g root dry weight	Range of nitrogen-fixing (acetylene-reduction) activity of undisturbed cereals-soil cores in nmol ethylene $core^{-1}$ 24 h^{-1}
$7.0x10^{1}-1.16x10^{4}$	$7.14x10^{2}-6.73x10^{5}$	20.6 - 29

The bacteria numbers are means of three different sample sites and two different wheat cultivars (Urban/Ralle) measured in 1986 and 1987.

different organic fertilizers (A and B) were slightly higher with organic manure A than with organic manure B and controls without organic fertilizer. This was possibly dependent on lower nitrogen content (0.9%) which did favour *Azospirillum* development, especially in earlier growth stages of wheat as shown in previous experiments (Markus and Krämer, 1987) (fig. 1). In inoculated variants with S1 and S12 always could be detected higher

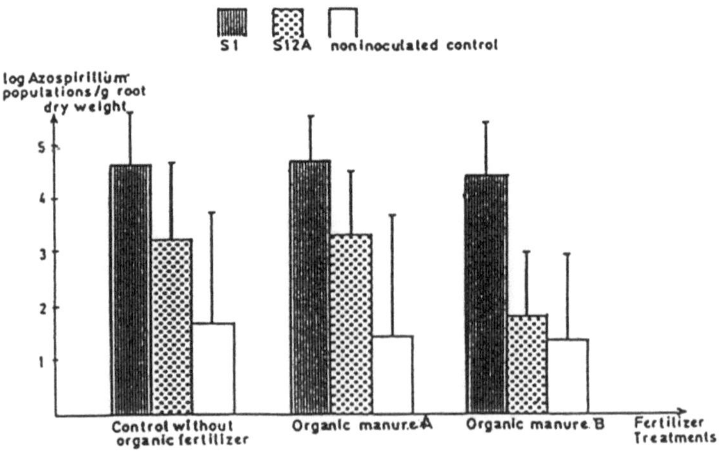

Fig. 1: Densities of Azospirillum spp. after inoculation with Azospirillum strains S1 and S12A in root environment of wheat (cv. Ralle) with two organic fertilizers. The numbers are means of all dates.

densities of *Azospirillum* spp. (fig. 1). These strains were proved to enhance plant and root development and grain yield (unpublished data). The use of organic manure to enhance *Azospirillum* populations and their activity in the rhizosphere of wheat had led to a high number of cellulytic fungi, streptomycetes and other bacteria (unpublished data). Therefore *Cellulomonas uda* was used in mixed culture with *Azospirillum* strain S1 to determine possible interactions.

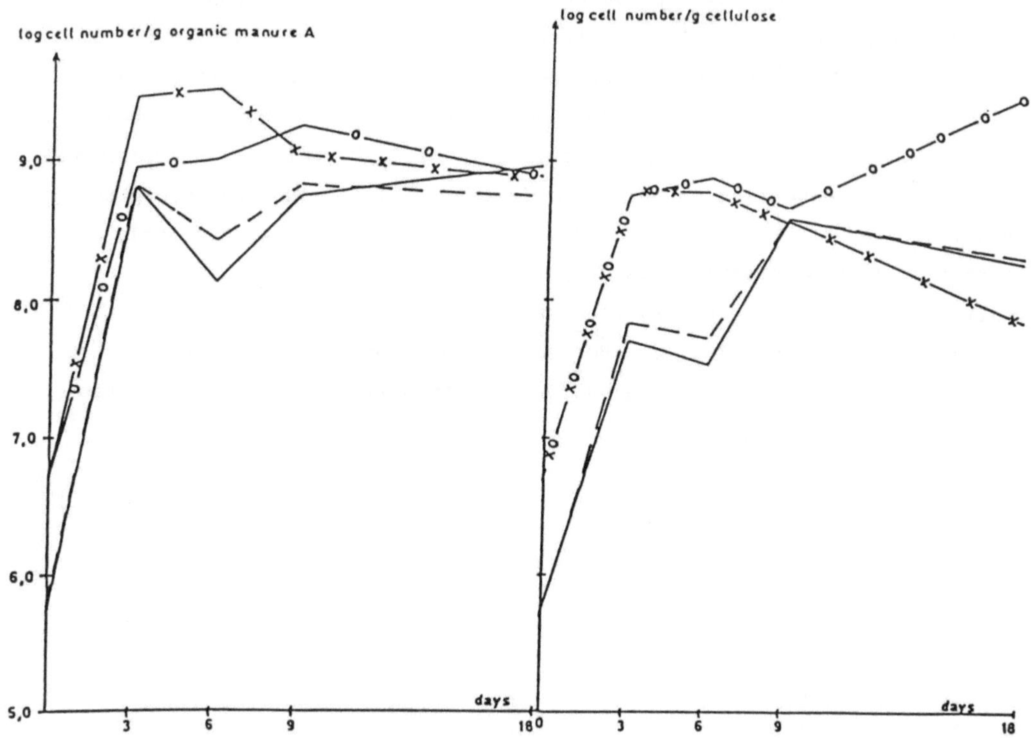

Fig. 2: Development of A. lipoferum S1 in pure culture and in mixed culture with C. uda on organic manure (left) and on cellulose (right).

———	A. lipoferum S1 x C. uda
- - - - -	A. lipoferum S1 in pure culture
—o—o	C. uda x A. lipoferum S1
—x—x	C. uda in pure culture

In 18-day-experiments with cellulose as a C-source. *C. uda* populations in mixed culture showed higher numbers beginning with the 6th day and increasing significantly to the 18th day (fig. 2, right) . *C. uda* in pure culture greatly decreased to log. 7.85. *C. uda* was favoured by *Azospirillum*-Co-culture possibly utilizing the fixed nitrogen after exhausting nitrogen sources in the medium. *Azospirillum* cells itselves used in mixed cultures formed organic compounds (organic acids etc.) by *C. uda* as indicated by decreasing pH-values in pure culture of *C. uda* and increasing pH-values in mixed culture with A. lipoferum S1 (fig. 3). *Azospirillum* was shown to use organic acids (e.g. malate, lactate) by Tarrand et al. (1978). With organic manure A bacterial numbers climbed to a higher level than the bacteria on cellulose (fig. 2, left). *A. lipoferum* S1 in pure culture decreased whereas *A. lipoferum* S1 in mixed culture with *C. uda* increased after 18 days probably using products of degraded organic material. This observation was supported by decreasing pH-value in pure culture of *C. uda* to 6.85 and increasing pH-value in mixed culture with *A. lipoferum* S1 to 8.03 (fig. 3). Between the 9th and 18th day higher populations of *C. uda* in mixed culture appeared than in the pure culture.

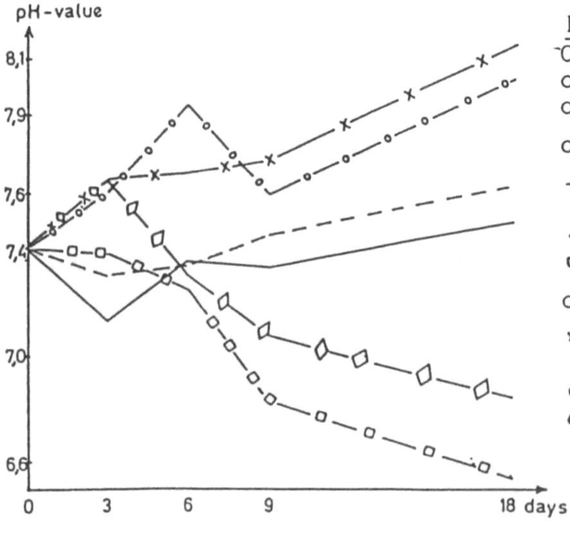

Fig.3: Course of pH-values in cultures of A. lipoferum S1 and C. uda on cellulose and organic manure

on cellulose:
- - - A. lipoferum S1 in pure culture
—— A. lipoferum S1 x C.uda
□—□— C. uda in pure culture

on organic manure:
×—×— A. lipoferum S1 in pure culture
o—o— A. lipoferum S1 x C.uda
◇—◇— C. uda in pure culture

Mixed cultures of *C. uda* and *A. lipoferum* S1 on cellulose and organic manure A were consistantly more effective in nitrogen fixation (acetylene reduction) than pure cultures (fig. 4). Nitrogen fixing activity on cellulose was markedly higher in mixed than in pure cultures of each strain. Hegazi et al. (1983) observed increasing acetylene reduction activity after straw amendments with maize, but no acceleration of *Azospirillum* numbers. The density of other native rhizosphere microorganisms increased. Cultures on organic manure declined faster after the 9th day in their activity than the cultures on cellulose, possibly due to the higher level of nitrogen in the organic manure A. This level did not demand nitrogen fixation of *A. lipoferum*-strain S1 like in nitrogen-defecient cellulose based medium. The low nitrogen content in cellulose treatments seemed to be responsible for marked differences in bacterial growth at the 18th day between mixed and pure cultures (fig. 2, right) and acetylene reduction activity. Similar results were obtained by Halsall and Goodchild (1986) with cultures of *A. brasilense* and *Cellulomonas* sp. grown on straw and cellulose.

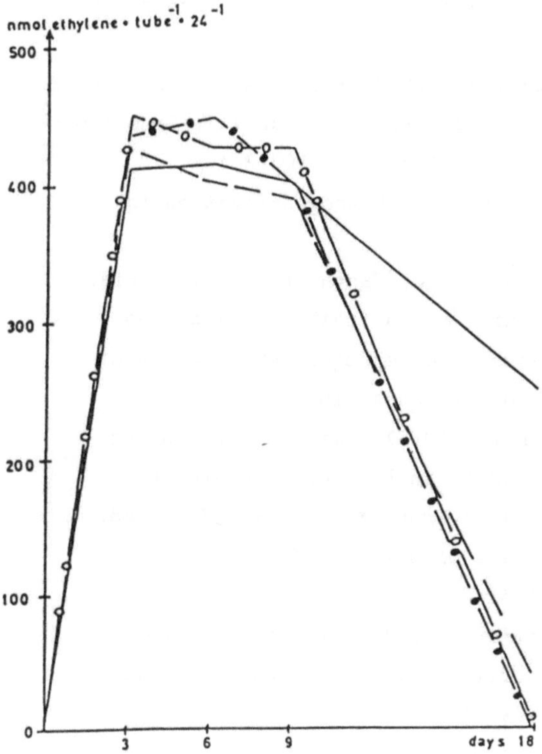

Fig.4:
Development of nitrogen-fixing (acetylene reduction) activitiy in pure and mixed cultures of A. lipoferum S1 and C. uda on organic manure and cellulose.

on cellulose:

——— A. lipoferum S1 x C.uda
- - - A. lipoferum S1 in pure culture

on organic manure:

o—o— A. lipoferum S1 x C.uda
•—•— A. lipoferum S1 in pure culture

This work has been supported by a grant of Wissenschaftsministerium Nordrhein-Westfalen, FRG.

References

Baldani, V.L.D. and J. Döbereiner (1979): Host-Plant Specifity in the Infection of Cereals with Azospirillum sp. Soil Biol.Biochem. 12, 433-439.

Gadkari, D. and H. Stolp (1974): Influence of Nitrogen Source on Growth and Nitrogenase Activity in Azotobacter vinelandii. Arch. Microbiol. 96, 135-144.

Halsall, D.M. and D.J. Goodchild (1986): Nitrogen fixation associated with development and localization of mixed populations of Cellulomonas sp. and Azospirillum brasilense grown on cellulose or wheat straw. Appl. Environ. Microbiol. 51, 849-854.

Hegazi, N.A. and M. Manib (1983): Response of maize plants to inoculation with azospirilla and (or) straw amendment in Egypt. Can. J. Microbiol. 29, 888-894.

Ishac, Y.Z., S.A.Z. Mahmoud, J. Krämer and W. Eweda (1982): Azotobacter Inoculation and Wheat Growth. Proc. 2nd Intern. Symp. on N_2-Fixation with Non-Legumes, Banff, Canada.

Jagnow, G. (1985): In Klingmüller, W. (ed.) Azospirillum III: Genetics, Physiology, Ecology. Springer-Verlag, Heidelberg, 203-214.

Markus, P. and J. Krämer (1987): Nonsymbiotic Nitrogen-fixation in Biological Farming Systems. Proc. IFOAM 6th Intern. Scient. Conference, Santa Cruz, California.

Nohrstaedt, H.J. (1984): Carbon monoxide as an inhibitor of N_2-ase activity (C_2H_2) in control measurements of endogenous formation of ethylene by forest soils. Soil. Biol. Biochem. 16, 19-22.

Okon, Y., S.L. Albrecht and R.H. Burris (1977): Methods for Growing Spirillum lipoferum and for Counting it in Pure Culture and in Association with Plants. Appl. and Environ. Microbiol. 33, 85-88.

Postgate, J.R. (1982): The fundamentals of nitrogen fixation. Cambridge University Press, 163-166.

Reents, H.J. (1987): Effect of different fertilizers on the potential N_2-fixing capacity (acetylene-reduction assay) of a sandy orthic luvisol. Z. Pflanzenernähr. Bodenk. 150, 31-36.

Tarrand, J.J., N.R. Krieg and J. Döbereiner (1978): A taxonomic study of the Spirillum lipoferum group, with descriptions of a new genus, Azospirillum gen. nov. and two species, Azospirillum lipoferum (Beijerinck) comb. nov. and Azospirillum brasilense sp. nov. Can. J. Microbiol. 24, 967-980 .

ASPECTS OF THE ECOLOGY OF AZOSPIRILLUM SP. IN BELGIAN SOILS

S. HOREMANS*, K. DE CONINCK AND K. VLASSAK

LABORATORY OF SOIL FERTILITY AND SOIL BIOLOGY, FACULTY OF AGRICULTURE, K.U. Leuven; Kard. Mercierlaan 92, B-3030 Leuven, Belgium

ACKNOWLEDGEMENTS : this study was supported by the "Instituut voor Wetenschappelijk Onderzoek in Nijverheid en Landbouw"; we had the skillfull assistance of S. Demarsin, the manuscript was typewritten by J. Pierre.

*Present address : S.E.S. - Industriepark 15 - 3300 TIENEN

SUMMARY

Azospirillum sp. was present in most of the Belgian fields. However it was never present in concentrations exceeding 10^3 cfu per gram of soil. Under field conditions none of the known species was found to be localised inside the roots of barley, wheat, corn or grasses. A. lipoferum was found only on corn and grass roots while A. brasilense was present under all crops investigated. When applied as a peat based seed inoculum A. brasilense established into the rhizosphere of field-grown winterbarley and winterwheat. The established population survived quantitatively the winterperiod. However, no indications for displacement or active growth during the springtime or summer were found.

Keywords : Soil Microbiology - Rhizosphere bacteria - Azospirillum sp. - seed inoculation - population dynamics.

INTRODUCTION

We investigated if Azospirillum sp. can be taken into account for use as a biological agent with industrial perspectives in temperate regions. More specifically we were looking for indications of possible activity on, or association with, the monocotyledonous crops of agricultural importance :

Azospirillum IV: Genetics, Physiology, Ecology
Edited by W. Klingmüller

winterwheat, winterbarley, rye, grasslands and corn. The study involved the estimation of frequency and absolute density of Azospirillum sp. in fields grown with these crops, the collection of data and the localisation of the bacteria in the rhizosphere system, the isolation and characterisation of endogenous strains, greenhouse experiments and field evaluation. In the course of the latter trials, the survival of the applied inoculum and its distribution within the rhizosphere were monitored as a function of time. Results on some of the ecological aspects of Azospirillum sp. will be discussed here.

MATERIALS AND METHODS

Media

A nitrogen free malate (NfM) stock solution was prepared, containing (per liter of distilled water) : malate, 50 g; KH_2PO_4, 6g; K_2HPO_4, 2g; $MgSO_4.7H_2O$, 2g;NACl, 1g; $Na_2MoO_4.2H_2O$, 20 mg; KOH, 49g; $FeCl_36H_2O$, 150 mg; $CaCl_22H_2O$, 250 mg; the final pH was adjusted to 6.8.
NfM semi-solid medium was prepared by diluting this stock solution in distilled water (1/9 v/v) adding 4 ml l^{-1} of a bromothymol blue solution (0,5% in ethanol), 100 mg of yeast extract and 2 $g.1^{-1}$ of agar. The agar was dissolved by boiling before autoclaving the medium. Culture medium was prepared by adding to the tenfold diluted NfM stock solution, 1 g l^{-1} of NH_4Cl. TAE buffer was a 50-fold dilution of a stock containing 242 g Trizma base (Sigma) and 37.2 g disodium EDTA per liter. The pH was adjusted to 7 with acetic acid and 7.5 g NaCl was added per liter. Phospate Buffer Saline (pH 7.2) was prepared by dissolving 1.15 g Na_2HPO_4, 0.2 g KH_2PO_4 and 7.5 g NaCl per liter of distilled water (PBS). All media were sterilised by autoclavation.

Strains used for inoculation

A mixture of four different endogenous strains of A. brasilense was used for inoculation. A. brasilense S631 was isolated by L. Reynders from a belgian soil. The other A. brasilense strains were isolated in the course of this project from winterwheat (strains 85W16 and 85W5) and corn (85M82) rhizosphere respectively. Based on 201 biochemical, nutritional and physiological tests a high degree of similarity with the type and reference strains and a perfect accordance with the species description was found (data not shown). Use of organic substances as a sole source of carbon and nitrogen and referred to in the results were done according to (1).

Sampling and counting procedure

If not stated otherwise the samples were taken between 0 and 15 cm deep. All samples were stored at 4°C up to analysis, maximally 24 hours later. Samples originating from different plots were stored separately. For studying the distribution in natural soils about 15 plants originating from minimally 5 different spots on a particular field, were sampled.

For monitoring the survival of the applied inoculum, soil not influenced by the roots ,"soil", was taken at 20 different sites between the plots. In these experiments 15 plants from each of the 6 treated plots were withdrawn. At the start of the counting procedure the soil hanging between the roots was removed by hand. This sample is defined as "rooted soil". From each of the 6 plots, 4 g of this rooted soil was pooled for analysis.

From the homogenised "soil" sample 24 g was analysed. The soil, and rooted soil samples were dissolved in 10 ml TAE buffer per g soil; the roots with firmly adhering soil were also taken up in minimally 10 ml of TAE buffer per gram of material. All 3 suspensions were shaken vigourously for 20 minutes and tenfold dilutions were made in PBS. The washed roots were either sterilised 4 min in ethanol and crushed than (survey study) or crushed immediately (survival study), shaken again for 20 min in TAE and diluted. This ethanol treatment did not kill effectively infecting, gramnegative non sporeforming bacteria (data not shown). The latter sample is defined as "roots" in both experiments. From each dilution 100 microliter of solution was transferred to 14 ml tubes containing 5 ml of NfM semisolid medium. Five tubes were used for each dilution. From 2 to maximally 14 days following transfer, acetylene reduction activity (ARA) was measured. If ARA was present a sample from the upper half cm of the medium was transferred to fresh NfM semisolid medium. Reappearance of the ARA together with the presence of a fine dense, white pellicle were the operational criteria for accepting the presence of Azospirillum sp. and the most probable numbers were determined accordingly.

Inoculation procedure

Peat was dried, milled, sieved (smaller than 0.1 mm), neutralised with $CaCO_3$ and sterilised. 125 ml of a 2 day old A. brasilense culture was added per 100 g of sterilised carrier material. After 1 week of incubation at 30°C the inoculum was stored at 4°C. Preliminar experiments indicated the A. brasilense peat cultures could be stored for up to 6 weeks at 4°C without any loss of viability. Just before planting, the seeds were coated using 10 g of inoculum and 4 g of arabic gum per 100 g of seeds. Seeds were desinfected with Sibutol[R] and Baytan[R] for winterbarley and winterwheat respectively. Growth of the applied bacteria was not retarded on solid medium containing per liter the equivalent of desinfecting material for 1 kg of seeds. However, higher doses were toxic.

RESULTS AND DISCUSSION

Field Survey

Frequency

According to the criteria given in material and methods Azospirillum sp. was present in 53 of the 65 soils analysed. No correlation with the type of the crop was evident. More specifically Azospirillum sp. was found in 16 out of 18 cornfields, 11 out of 13 winterwheatfields, 12 out of 16 winterbarleyfields, 1 out of 2 ryefields and 13 out of 16 grasslands. Moreover Azospirillum sp. was found in all types of soil texture analysed. It was present in 10 out of 14 sandy soils, in 4 out of 6 sandy loam soils, in 21 out of 24 clay soils, in 3 out of 3 calcarous soils and in 16 out of 18 loamy soils. We can conclude that Azospirillum sp. is widely distributed in belgian agricultural soils.

Absolute densities

We estimated absolute densities of Azospirillum sp. separately in the rhizosphere soil (including those bacteria that can be washed free from the rootsurface), those in the soil not in direct contact with the roots and

Table 1 : Incidence of *Azospirillum* sp. at different field locations under winterwheat cultivation as evaluated by MPN counts

Field site	Soil texture	Number of *Azospirillum* sp. /gr rooted soil	/gr rhizosphere soil	/gr fresh root
Bertem	sandy loam	0	0	0
Tiegem	sandy loam	8	82	0
Waremme	loam	93	103	0
Branchon	loam	15	7	0
Thorembais-les Béguines	loam	75	10	0
St.Truiden 1	loam	24	10	0
St.Truiden 2	loam	15	60	0
Gistel	clay	43	232	0
Stuyveskerke	clay	240	1163	0
Houtave	clay	0	0	0
St.Pieters	clay	23	34	0
Klemskerke	clay	4	244	0
Eggewaert	clay	7	24	0

these within the roots. This was done for winterwheat (illustrated in table I), rye, winterbarley and for grasses . For corn only frequencies have been scored.
Azospirillum was never detected within surface sterilised roots of corn, rye, winterbarley or winterwheat. However in one of the 13 grassland samples where Azospirillum sp. was present in the soil, Azospirillum sp. activity was also found within the roots.
In the rhizosphere soil and in the soil further from the roots the numbers of Azospirillum sp. per gram soil ranged from 3 to maximally 1163. Similar densities were found in France (J. Balandreau, pers. communication) and in Germany (P. Markus and J. Krämer, this volume). Other bacteria, also growing in NfM semi-solid medium were present in densities from 10.000 to over a million times higher (data not shown).
In 17 out of 37 samples the most probable number of Azospirillum sp. was not significantly higher in the soil further from the roots as compared to the rhizosphere soil. At least in these cases no enrichment on the root surface occurred. In 20 out of 37 samples the numbers in the rhizosphere soil were higher. However, due to the heterogenous nature of the rhizosphere sample (soil + rootsurface) we can not draw the conclusion that rhizosphere enrichment indeed occurred here.

Identification of the diazotrophs

From these samples we identified a total of 28 diazotrophs : 12 from corn, 7 from winterwheat, 3 from barley, 5 from grasses and one from rye. Based on 201 physiological, biochemical and nutritional characteristics, 8 were identified as A. lipoferum, 18 as A. brasilense and one strain had intermediate characteristics. One strain represented a for us unknown diazotrophic species. It was the strain responsible for the Azospirillum like activity found within surface sterilised grassroots. In general, A. lipoferum grew much better on amino acids and amino acid derivatives. More specifically, under the conditions described under material and methods, the strains classified as A. lipoferum grew all on suberate, azelate, adipate, pimelate, sarcosine, ethanolamine, L-serine, phenylalanine and tyrosine. Other products were used by almost all A. lipoferum strains and not by A. brasilense (e.g. arginine, threonine, 2-ketoglutarate, D-alanine, levulinate). Histidine was used by all A. lipoferum strains and but by one

A. brasilense strain. On asparate and citrate all A. lipoferum strains grew vigorously, A. brasilense in contrast grew very slowly. All of the A. brasilense could use itaconate, mesaconate and xylose. The former 2 products were not used by A. lipoferum and xylose only by half of the A. lipoferum strains. A. brasilense was recovered from all 5 crops investigated. A. lipoferum could only be isolated from cornfields and grasslands indicating a certain degree of selective pressure exerted by the hostplant.

Inoculation studies

The suitability of a peatbased seedinoculum for applying A. brasilense

From the seeds treated as described sub material and methods between twentythousand and one million of living bacteria per seed (see table 2) were recovered at the moment of planting.

Table 2 : Numbers of Azospirillum sp. recovered per seed after a peat-based seed inoculation as evaluated by MPN counts.

crop	field site	date of inoculation	date of counting procedure	Number of Azospirillum sp./seed
winterbarley	Tienen	01/10/85	07/10/85	2.6×10^4
	Tiegem	30/09/85	07/10/85	4.5×10^4
winterwheat	Tienen	04/11/85	05/11/85	85.7×10^4
	Tiegem	25/10/85	28/10/85	29.8×10^4

Establishment of the inoculum

The experiment was performed at two field sites : one loamy soil and one sandy loam. At each location both winterbarley and winterwheat were treated. A randomised block design was followed, and 6 repetitions were used. The actual data for one of these experiments are given in table 3. As can be seen in this table and confirmed in all 4 experiments the numbers of A. brasilense recovered in the rhizosphere are much higher for the inoculated plots as compared to the control plots. The control plots contained only very low numbers of bacteria, confirming the earlier survey and extending these data over the whole growing season. The actual numbers counted at the moment of tillering in the rhizosphere of inoculated plots ranged from 1000 to 50 000 living cells per gram of rhizosphere soil. The established inoculum survived quantitatively the winter period. In all soil compartments the actual numbers were identical before and after the winter, both in the loamy soil and in the sandy loam soil (table 3). However, during the following spring, no increase in numbers was observed. In contrast, in all 4 experiments densities of Azospirillum sp. decreased slowly to reach a level between 30 and 1000 living cells per gram of rhizosphere soil or per gram of roots. For characteristics Azospirillum does share with other rhizo-bacteria, more specifically the production of auxins and cytokinins (2,3) one must accept that bacterial densities as found in this study are largely insufficient to have any determining effect on plant performance.

Dispersion of the inoculum

The inoculum was found in only very low densities in the soil taken between the roots. This indicates that either the inoculated cells did not migrate away from the roots or that they did not survive in this environment.
A small part of the A. brasilense population consistently associated with the washed and the crushed roots. However the highest densities were recovered in the rhizosphere sample.
When the distribution of the inoculum was followed as a function of depth we recovered the complete Azospirillum sp. population in the upper 10 cm of the root system (table 4). These observations suggest the abscence of active displacement of the applied inoculum along the growing root.

Table 3 : Numbers of Azospirillum sp. in different soil compartments after seed inoculation of winterbarley with a mixture of 85M82, 85W5, 85W16 and S631. Data evaluated by MPN counts are from the field site in Tienen (loamy soil)

Treatment	Sample	Numbers of Azospirillum sp.* at different sampling dates.			
		21/12/85	11/03/86	26/05/86	26/06/86
Non-inoculated	soil	2	2	4	0
	rooted soil	2	20	70	0
	rhizosphere soil + rootsurface	81	62	183	3
	roots	68	0	15	0
Inoculated	rooted soil	220	330	93	39
	rhizosphere soil + rootsurface	1.1×10^5	6.5×10^4	1.5×10^4	5.0×10^2
	roots	2.9×10^3	2.05×10^3	6.0×10^2	1.8×10^3

* Numbers of Azospirillum spp. are expressed per gram of soil for the "soil" samples; for the other counts results are expressed per gram of fresh root.

Table 4 : Incidence of A. brasilense in the rhizosphere soil of winterbarley as a function of depth after seed inoculation with 85M82, 85W5, 85W16 and S631. Data, evaluated by MPN counts, are from a field experiment in Tienen

Depth	Roots fresh weight as a % of total root weight in 0-80 cm	Number of Azospirillum brasilense per gram soil
0-10 cm	72 %	1.7×10^3
10-30 cm	21 %	0
30-60 cm	5.4 %	0
60-80 cm	1.8 %	0

LITTERATURE

(1) Kersters, K.; Hinz, K.H.; Hertle, A.; Segers, P.; Lievens, A.; Siegman, O. and De Ley, J. (1984); Int. J. Syst. Bacterial 34 : 57-70.

(2) Horemans, S. and Vlassak, K. Production of indol-3-acetic acid by Azospirillum brasilense (1985).
In Klingmüller, W. ed. Azospirillum III : Genetics, physiology, Ecology. Springer Verlag, Heidelberg, p. 98-108.

(3) Horemans, S.; De Coninck, K.; Neuray,J.; Hermans, R. and Vlassak, K.;
Production of Plant Growth Substances by Azospirillum sp. and other Rhizospere Bacteria (1986).
Symbiosis, 2 pp. 341-346 .

MODIFICATION OF SOIL ENVIRONMENT THROUGH STRAW APPLICATION VERSUS AZOSPIRILLIM spp. INOCULATION

N. A. Hegazi
Faculty of Agriculture, Cairo University, Giza, Egypt

Summary: Effect of straw amendment on development of wheat plants inoculated with *Azospirillum* spp. and/or N-fertilized was investigated under field conditions of arid zones. Irrespect of any fertilization treatments, incorporation of straw into soil supported better growth of plants. Grain and dry matter yields increased with 8-12 % and 15-37 %, depending on fertilization treatments, respectively in absence of straw. Such increases amounted to 25-47 % and 76-146 % in presence of straw. Inoculation with azospirilla as such did satisfy 30 % of the N-requirements of plants, and straw amendment boosted effects of inoculation particularly in presence of moderate doses of N-fertilization.

Introduction

Soils of the arid zone (Aridisol) -in particular sandy soils- are known to be poor in organic matter. Generally, they possess low numbers of soil bacteria including dinitrogen fixers (1, 5, 10). Cultivation of such soils requires adoption of a comprehensive fertilizing regime. Application of minimum tillage and incorporation of plant refuses build up soil organic matter and modify soil environment to achieve maximum productivity. Our previous experiments with incorporation of wheat and maize straw into soil demonstrated positive effects on ATP generation, population densities and nitrogenase activity of diazotrophs in soil (8, 11). With pot experiments, response of sorghum seedlings to azospirilla-inoculation was significantly improved in soil amended with 2% wheat straw (4). As a continuation of this work and of previous inoculation trials (5, 10), the present study reports on field responses of wheat, grown in sandy soils, to straw amendment and/or azospirilla inoculation under arid conditions.

Azospirillum IV: Genetics, Physiology, Ecology
Edited by W. Klingmüller

Materials and methods

Inocula:

Three isolates of Azospirillum brasilense, previously isolated from Sudan grass (10) were used. They were grown independently in semi-solid malate medium (7) for 4 days. Mixed inocula were prepared by mixing equal portions (v/v) of bacterial suspensions (2 x 10^8 cells ml^{-1}). Seeds were soaked in the liquid inocula for ca. 20 minutes and were left to dry overnight. They possessed 36 x 10^4 cells $seed^{-1}$.

Field experiments:

The following experiment was executed in sandy-loam soils of Gassim, Saudi Arabia (10). A split-plot design in a randomized complete block with 3 replicates was used. Straw amendment was assigned to the main plots. There were 6 fertilization treatments: 1) Control; 2) azospirilla inoculation; 3) full dose of N only (200 kg N ha^{-1}); 4) 1/3 dose of N only (67 kg N ha^{-1}); 5) simultaneous azospirilla inoculation + full dose of N; 6) simultaneous azospirilla inoculation + 1/3 dose of N. Treatments were distributed randomly in sub-plots. In case of straw application, 2 kg of fine wheat straw was added and ploughed into the top 15 cm of each sub-plot. Phosphorous (calcium triple phosphate 45-47 % P_2O_5) per hectare was carried out during seed bed preparation. N-fertilization, in the form of urea 45 %, was applied in 3 doses, at planting, and after 4 and 8 weeks. The sub-plot was 4.8 m^2 consists of 8 rows 4 m long and 15 cm apart. Seeds of cv. Yoco-Rogo were sown on 15.1.1985 and were sprayed, before covering, with suspension of azospirilla (5 x 10^6 cells ml^{-1}) either alive (for inoculated treatments) or autoclaved (for all others). General agronomic practices were employed. Samples, of minimum 10-30 plants, were withdrawn periodically from the outer rows for the determination of plant height, total dry weight of roots, shoots and ears as well as numbers of azospirilla in free soil and on roots, i. e. histosphere (10). At harvest, dry matter and grain yields were obtained for each sub-plot. Organic carbon and total nitrogen were determined for soil and plant samples (3).

Statistical analysis

Results were analysed statistically using the analysis of variance.

Results and discussion

Incorporation of straw into soil exerted positive effects on growth of wheat plants. Irrespect of fertilization treatments, plants were much taller ($p = 0.01$,Fig. 1) and scored higher dry matter of shoots due to straw amendment. Such significant effects were much more pronounced for N-fertilized plants with or without azospirilla inoculation. Dry weight of roots, of plants up to 9 weeks-old, did not respond similarly (Fig. 2) until a later stage, 12 - 13 weeks-old plants, when a rather significant increase ($p = 0.01$, Fig.3) in root as well as shoot weight was quite obvious. This delayed positive effect of straw amendment is understood as straw decomposition by microbial community of the soil and the release of nutritive by-products elapse for sometime (8,11), being weeks in soils under investigation (Fig. 4).

Root development was affected due to fertilization treatments as such, and a significant increase in root biomass was particularly recorded when a limited dose of N was given simultaneously with azospirilla inoculation ($p = o.01$). An effect that was previously established for wheat and other cereal crops (2, 10, 12, 13, 15). The application of straw, irrespect of fertilization, magnified the increase in root biomass which corresponded to 130 - 568 % compared to untreated plants ($p = 0.01$). On the other hand, inoculation together with high dose of N diminished such positive effect which might be considered as an ecological side-effect of antagonistic relationship arises among specific groups of soil bacteria, flourish under these conditions, and their by-products from onde side and the azospirilla population and plant roots on the other side. Azospirilla were found with particular dense populations ($> 5 \times 10^6$ plant root^{-1}) in histosphere of plants-grown in presence of straw - of all fertilization treatments except those receiving high dose of N accompanied with azospirilla inoculation (Table 1). Inoculation as such resulted in 9.1 and 128.8 % increases in azospirilla densities found in histosphere of plants grown in absence and presence of straw respectively. Straw application increased as well the total microbial population in soil (2 - 7 folds, Table 1).

Dry weight of ears followed a similar trend and significant effects were attributed to straw amendment ($p = 0.01$) and fertilization treatments ($p=0.05$).

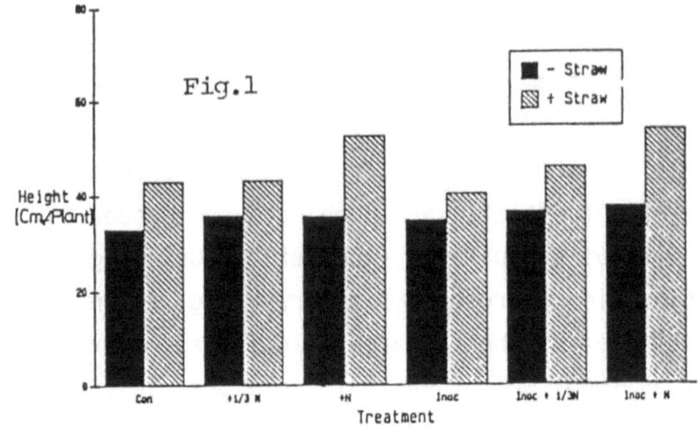

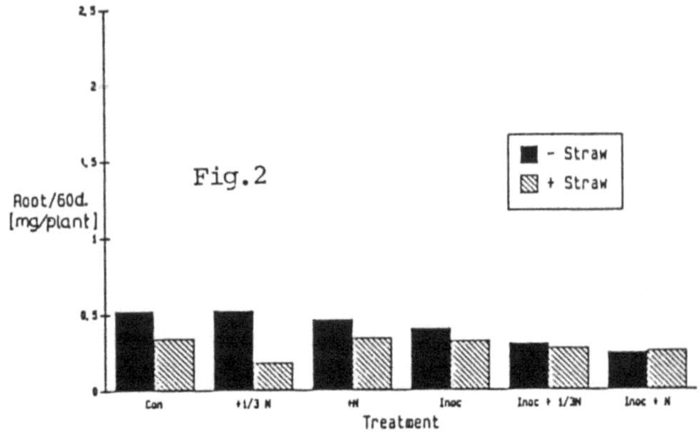

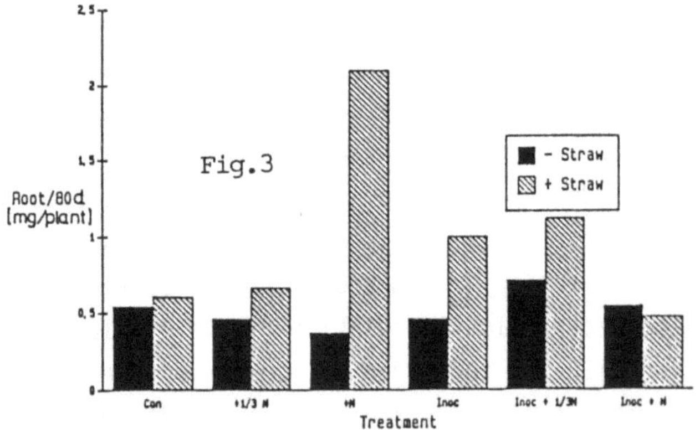

Effects of straw amendment, inoculation and N-fertilization on height of plants (Fig.1) and root biomass for 60 days-(Fig.2) and 80 days-(Fig.3) old plants.

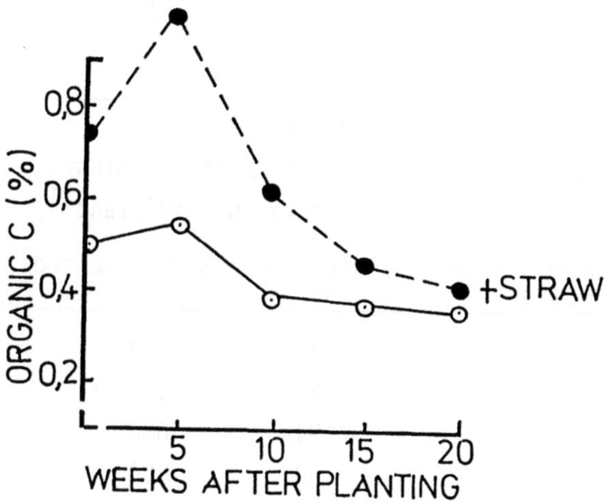

Fig.4-Organic carbon decomposition in soil amended (●–●) and not amended (○–○) with straw.

At harvest, significant differences in dry matter yield ($p = 0.01$) were distinguished among fertilization treatments as well as straw amendments (Table 2). Irrespect of straw amendment, yield of azospirilla-inoculated plants and those fertilized with a low dose of N (67 kg N ha^{-1}) were comparable. This means that inoculation did satisfy ca. 30% of the N- requirement of growing plants (5, 10, 13, 14, 15). While higher yields were obtained with full N- fertilized plants, best records were obtained for those plants receiving simultaneous full N-fertilization and azospirilla inoculation. These significant results apllies also to the total N-yield taking into consideration the N-content detected for plants at harvest (0.4 - 0.77 %) which did not vary significantly.

Grain yield (Fig. 5) differed significantly ($p = 0.01$) among fertilization treatments, straw amendment and 2-way interactions ($p = 0.05$). In presence of straw, inoculation resulted in grain yield increases of 31%, and of 25 and 45% in presence of low and high dose of N respectively. Without straw amendment, increases were only 9, 8 and 12% in the same respective order. Grain yield but not dry matter yield of non-treated plants, i.e. non-fertilized, grown in straw-amended soil were lower to those of unamended soil. This might be due to lack of N in soil and consequent active competition among flourishing soil microflora and plant roots to scavenge available N under these particular conditions.

Table 1 - M.P.N. estimates of Azospirillum in soil [a] and histosphere of wheat plants

Treatments	m.p.n. estimates of Azospirilla in Free soil ($x10^5 g^{-1}$)		Histosphere of plants grown in ($x10^4 plant^{-1}$)	
	unamended soil	amended with straw	unamended soil	Straw-amended soil
untreated	13	70	44	440
+1/3 dose N	35	46	66	560
+ Full dose N	11	33	34	1080
Azospirilla-inoculated	33	49	98	700
Inoculation + 1/3 N	35	110	66	480
Inoculation + Full N	35	33	158	340

a Total bacterial countes were $12\text{-}25x10^6$ and $5\text{-}20x10^7 g^{-1}$ free soil of un-amended and straw-amended soil respectively

Table 2 - Effects of straw-amendment and various fertilization treatments on total dry matter yield

Treatments	Dry matter yield (g $plot^{-1}$)		
	without straw-amendment	with straw-amendment	Mean
untreated	3338.9	4949.7	4144.3
1/3 dose-N	4093.3	5876.6	4984.9
Full dose-N	4033.8	8274.0	6153.9
Azospirilla-inoculation	3468.9	6175.1	4821.9
Inoculation + 1/3 N	3855.0	7483.7	5669.4
Inoculation + Full N	4527.7	7374.1	5950.9
Mean	3886.3	6688.9	

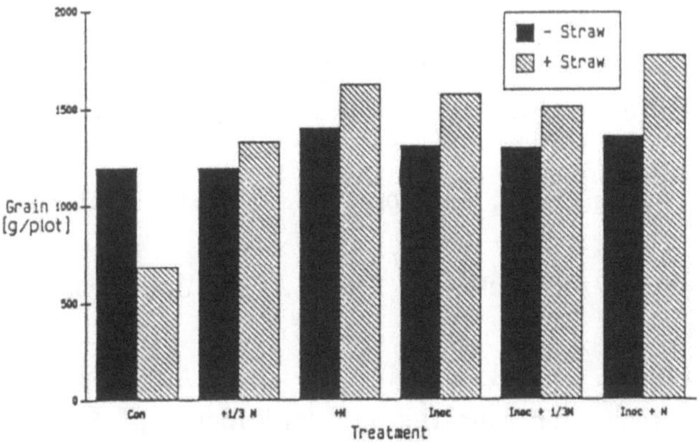

Fig.5-Grain yield as affected by straw amendment, inoculation and fertilization treatments.

Acknowledgements

This work was supported by College of Agriculture and Veterinary Medicine, King Saud University at Gassim, Saudi Arabia. Thanks are due to H. Saleh and A. Ismael for active participation and cooperation, and to A. Hussein, O. Abou-Geab, M. El-Dochael and S. Al-Sahw for excellent technical assistance. Statistical analysis was carried out with the collaboration of R. Brandl and, H. El-Khawas of Bayreuth university.

References

1. Abd-el-Malek, Y. (1971). Plant and Soil. Special volume, 423-442.
2. Baldani, V., Baldani, D., Ivo Baldani, J. and Döbereiner, J. (1983). Can. J. Microbiol. 29, 924-929.
3. Bremner, J.M. (1965). In Methods of soil analysis. II. Chemical and microbiological properties. American Society of Agronomy, Madison, WI. P. 1149-1178
4. Eid, M., Monib, M., Jensen, V. and Hegazi, N.A. (1983). In Strub, A., Chartier, P. and Schleser, G. (eds.). Energy from Biomass. 2nd E.C. Conference. Applied Science Publishers, Essex, p. 205-209.
5. Hegazi, N. A. (1983). In Klingmüller, W. (ed.) Azospirillum II: Genetics, Physiology, Ecology. Experienta Supplementum 48, Birkhäuser-Verlag, Basel, p. 171-189.

6. Hegazi, N. A. and Al-Sahael, Y. (1983). Egyptian Society of Applied Microbiology. Proceedings of V Conf. Microbiol., Cairo, May 1983. p. 49-63.
7. Hegazi, N. A., Amer, H. and Monib, M. (1979). Soil Biol. Biochem. 11, 437-438.
8. Hegazi, N. A., Khawas, H. M., Farag, R. S. and Monib, M. (1986). Plant and Soil 90, 383-389.
9. Hegazi, N. A., Monib, M., Amer, H. and Shokr, E. (1983). Can. J. Microbiol. 29, 888-894.
10. Hegazi, N. A. and Saleh, H. (1985). In: Klingmüller, W. (ed.) Azospirillum III: Genetics, Physiology, Ecology. Springer-Verlag, Heidelberg, p. 189-202.
11. Hegazi, N. A., Vlassak, K. and Monib, M. (1979). Plant and Soil, 51, 27-37.
12. Jain, D.K. and Patriquin, D.G. (1985). Can. J. Microbiol. 31, 206-210.
13. Kapulnik, Y., Sarig, S., Nur, I., Okon, Y., Kigel, J. and Henis, Y.H. (1981). Expl. Agric. 17, 179-187.
14. Okon, Y. (1984). In: Advances in nitrogen fixation research. Proceedings V Int. Symp. on Nitrogen Fixation. Noordwijkerhout, the Netherlands. Marinus, Nijhoff/Dr. W. Junk Publishere, p. 303-309.
15. Okon, Y. and Kapulnik, Y. (1986). Plant and Soil 90, 3-16.

AZOSPIRILLUM SPP ECOLOGY
OF SOME SOILS OF THE SOMALI REPUBLIC

F. Favilli, F. Trinci and W. Balloni
Istituto di Microbiologia Agraria
e Tecnica Universita' degli Studi di Firenze, Italia

Summary: During a survey of the distribution of nitrogen-fixing bacteria in soil samples (12 alluvial and 1 sandy) collected in two localities (Jenale and Afgoj) in the Somali Republic, microorganisms resembling *Azospirillum* were isolated and characterized.

All the soil samples exhibited weak nitrogenase activity when incubated in presence only of sodium malate. *Azospirillum*-like bacteria occurred only in the 12 alluvial soil samples and not in the sandy one.

From the 4 soil samples, with the highest nitrogenase activity 18 *Azospirillum* strains were isolated altogether, and these were preliminarily identified as: *Azospirillum brasilense* nir+ (13 strains), *Azospirillum brasilense* nir- (1 strains).

Four strains showing physiological characteristics of the species *Azospirillum lipoferum* did not fit completety the description of the species and were considered very close to *Azospirillum lipoferum*.

The Azospirilla detected in these soils represented a considerable part (30-40%) of the entire diazotrophic population; in which *Azotobacter* sp., *Beijerinckia* sp. and heterocistous cyanobacteria (*Nostoc* sp., *Calothrix* sp., *Anabaena* sp. and *Scytonema* sp.) occured.

The collected data also represent the first contribution to our

Azospirillum IV: Genetics, Physiology, Ecology
Edited by W. Klingmüller

knowledge of the ecology of Azospirillum spp. in the soils of the Somali Republic.

Keywords: Azospirillum, occurrence, tropical soils.

Introduction

Nitrogen-fixing bacteria of the genus Azospirillum have been reported to occur in soils and in association with the roots of several plants widely distributed in tropical and temperate regions (1-2 - 3-4).
Moreover, the high incidence of Azospirillum in soils and in root samples from tropical regions relative to those from temperate zones can be attributed either to the low level of available nitrogen in tropical soils or to the higher temperature requirement of these microrganisms (5-6-7).
In a survey of the distribution of nitrogen fixing bacteria in Somalian soils, microorganisms resembling Azospirillum were isolated and characterized.

1 Material and Methods

1.1 Samples

The soil samples (12 alluvial and 1 sandy, each uncultivated) were collected at the end of the dry season in 1985 in several localities in the districts of Jenale and Afgoj in the Somali Republic.
The collected soil samples were stored in polyethylene bags at 10°C for two weeks, then transported to the laboratory (in Italy), where they were examined for the occurrence of nitrogen-fixing microrganisms with particular attention to the presence

of *Azospirillum*.

1.2 Nitrogenase activity of the soils

Ten grams of each soil sample either without C source or with I% of sodium malate were put into serum bottles (28 ml) and I ml of distilled water was added to each bottle.
The bottles were capped and preincubated for 16 h at 33°C in an atmosphere of N_2: O_2 (99:I v/v). After incubation the atmosphere was replaced with a mixture of air and acetylene (87.5:12.5 v/v) and the bottles incubated again for 1 h.
Ethylene was assayed by gas chromatography and nitrogenase activity was expressed as n.moles of C_2H_4 produced per gram (dry weight) of soil per hour.

1.3 Isolation and characterization of Azospirillum from soil samples

Enrichment cultures of *Azospirillum* were prepared by introducing 0.1 g of soil to 20 ml test tubes containing 5 ml of N-free semisolid malate medium (8). After 48 h of incubation at 33°C the tubes showing white subsurface pellicle, nitrogenase activity more than 10 n.moles ethylene ml^{-1} h^{-1} as well the presence of very active bacteria with the typical spiral shape and ßpoly-hydroxybutyrate (PHB) granules were considered as positive for the occurrence of *Azospirillum*-like forms.
Isolates of N_2-fixing spirilla were purified by several subcultures on Nfb supplemented with Congo Red according to the method described by Rodrigues-Caceres (9).
The identification of the isolates was made following the criteria suggested by Tarrand et al. (10), Baldani and Dobereiner (11) and Dobereiner (12).

Azospirillum lipoferum strain A3a from grass (Senegal), *Azospirillum brasilense* SP7 from *Digitaria decumbens* (Brasil) and *Azospirillum brasilense* strain cd from *Cynodon dactylon* (California) were used as reference strains.

The CFU of *Azospirillum* per gram (d.w.) of soil was estimated by the MPN method (13).

Nitrogen fixation of pure cultures of *Azospirillum* was tested by inoculating 0.2 ml of 48-h-old culture (grown in N-free semisolid medium) into 20 ml tubes with sodium malate and glucose as C sources.

The cultures were incubated for 48 h at 33°C. After incubation twelve per cent of the gas phase was replaced by acetylene and nitrogenase activity was expressed as n.moles of C_2H_4 per milligram of protein per hour. The protein content of each culture was determined from three replicates by using a modified Lowry method (14)

2 Results and discussion

Each of the soil samples exhibited nitrogenase activity in the presence of I% sodium malate though at different levels only (Tab.1).

In the majority of soil samples the levels of nitrogenase activity were between 18 and 76 n.moles C_2H_4 h^{-1} g^{-1} (d.w.) while in the samples AI, WBI, SH2 and SH5 the values were lower than 2 n.moles C_2H_4 h^{-1} g^{-1} (d.w.).

The enrichment cultures yielded *Azospirillum*-like forms only in the 12 alluvial soils and not in the sandy one (Table 2).

It was also possible to observe a direct correlation of the occurrence of *Azospirillum*-like forms with the nitrogenase activity of the enrichment cultures and the number of azospirilla of the soil examined.

Table 1 - Nitrogenase activity and characteristics of the soil samples.

Samples	Localities	Soil	pH	n.moles $C_2h_4h^{-1}$ g^{-1} (d.w.) (a) (soil amended with 1% sodium malate)
WVI	OWDHEGLE (JENALE)	Alluvial	8.11	18.0
WV2	"	"	8.00	41.0
WV3	"	"	7.55	28.0
WV4	"	"	7.66	35.0
WBI	"	"	8.00	0.15
SHI	SHAAM (JENALE)	"	7.87	76.0
SH2	"	"	7.70	0.6
SH3	"	"	8.00	19.0
SH4	"	"	8.00	36.0
SH5	"	"	7.72	0.16
MI	MORDINLE (AFGOJ)	"	8.00	53.0
M2	"	"	8.01	18.0
AI	LAFOOLE (AFGOJ)	Sandy	7.51	1.6

(a)average of two experiments, each carried out with three replicates

The nitrogenase activity of the sample Al (sandy soil) can be ascribed to other microorganisms belonging to other N_2-fixing species. In fact the azospirilla detected in these soils represented a considerable proportion (30-40%) of the entire diazotrophic population, in which Azotobacter sp., Beijeninckia sp. (Favilli unpublished data) and heterocistous cyanobacteria (15) of the genus Nostoc sp., Calothrix sp., Anabaena sp. and

Scytonema sp. occurred as well. From four enrichment cultures showing higher levels of nitrogenase activity eighteen strains of *Azospirillum* were isolated altogether.

Table 2. Occurrence of Azospirillum-like forms, nitrogenase activity in the enrichment cultures and CFU of azospirilla in the soil examined.

Samples	Subsurface pellicle	*Azospirillum*-like forms (a)	n moles C_2H_4 h^{-1} g^{-1} (d.w.) (b)	CFU 10^3 g^{-1} (d.w.)
WVI	+	+	38.74	420
WV2	+	++	61.00	560
WV3	+	+	48.65	450
WV4	+	++	55.40	540
WBI	+	+-	5.22	2.1
SHI	+	++	96.98	625
SH2	+	+-	6.95	1.00
SH3	+	+	39.66	250
SH4	+	+	56.77	15.0
SH5	+	+-	6.70	2.6
MI	+	++	73.97	570
M2	+	+	38.50	412
AI	-	-	6.37	-

(a) ++ elevated occurrence of Azospirillum-like forms
(b) average of two experiments, each with three replicates

Table 3 - Strain isolated from four soil sample enrichment cultures

Soil samples	n° of strain isolated	Isolate designation
WV2	6	1W; 2W; 3W; 4W; 5W; 6W;
WV4	2	7W; 8W
SH1	6	1S; 2S; 3S; 4S; 5S; 6S.
M1	4	1M; 2M; 3M; 4M.

Table 3 deals with the soil sample enrichment cultures from which the strains were isolated and the isolate designation.

2.1 Morphological, cultural and physiological characteristics of the strains isolated

After 24h in NFb semisolid malate medium the cells of the strains isolated from the four enrichment cultures were curved, highly motile rods with spiral movements. They contained PHB granules, and the size of the cells ranged from (0.9-1.4)x(2.2-3.5) µm.

After 48h in alkaline NFb semisolid malate medium supplemented with NH_4NO_3 (5mM) the cells of the strains 1W, 3W, 5W and 6W varied in length from 3.5 to 5.0 µm without PHB granules.

On NFb malate agar (12.5 g l^{-1}) with yeast extract (0.1 gl^{-1}) the colonies were white, round and flat, becoming pink after 15 days.

After 24h on PSS agar (16) the colonies appeared white and creamy, becoming light red after 25-30 days.

On NFb semisolid malate medium all the strains developed a heavy white pellicle that became pink after 15-20 days.

The strains grew well on organic acids as malate and succinate while α-Ketoglutaric acid, glucose, ribose and mannitol were utilized only by the strains 1W, 3W, 5W and 6W. The same strains utilized glucose, as C-source for nitrogenase activity with slight acidification of the medium (Table 4).

All the strains except strain 1S reduced nitrate to nitrite and nitrite to nitrous oxide or nitrogen gas.

All the strains were catalase positive and grew well without biotine.

Table 5 summarizes the morphological and physiological characteristics of the strains.

Table 4 - Nitrogenase activity of the strains during growth on malic acid and glucose as C sources.
48h-old culture on (NFb semisolid medium, p O_2 0.001, pH 6.8)

STRAINS	n.moles $C_2H_4h^{-1}$ g^{-1} protein (a)			
	malic acid	pH	glucose	pH
1W	39.34	7.7	13.79	6.4
2W	6.66	8.0	NG	
3W	22.15	7.8	6.76	6.5
4W	7.76	7.8	5.98	6.3
5W	16.05	7.7	5.20	6.3
6W	71.67	7.8	NG(°)	-
7W	9.65	7.8	NG	-
8W	48.13	7.8	NG	-
1S	64.82	7.8	NG	-
2S	69.14	7.8	NG	-
3S	4.23	8.0	NG	-
4S	9.75	7.9	NG	-
5S	11.31	7.7	NG	-
6S	17.86	7.8	NG	-
M1	8.95	7.8	NG	-
M2	9.75	7.9	NG	-
M3	7.25	7.9	NG	-
M4	8.76	8.0	NG	-

(a) means of two experiments, each carried out in three replicates.
(°) NG no growth

Table 5 - Morphological and physiological characteristics of the isolated strains

Characteristic	n° of positive strains
Glucose as sole C source	4
Biotine requirement	0
PHB in semisolid malate medium	18
Gas from nitrite	17
Catalase	18
Motility	18
Cell length from 3.5 to 5.0 µm	4
Cells shorter than 3.5 µm	14

Two groups of isolates were distinguished. One, including the majority of the strains, fitted the description according morphology and physiology of Azospirillum brasilense, while the characteristics of the other group of isolates (strains 1W, 3W, 4W and 5W) suggest their possible placement within the species of Azospirillum lipoferum even if they did not fit completely the description as concerns biotine requirement and absence of the typical elongated cells.

All the isolated strains can be temporarily classified as summarized in Table 6.

Table 6 - Proposed systematic position of the isolates

Species	n° of strains
Azospirillum brasilense nir+	13
Azospirillum brasilense nir-	1
Azospirillum lipoferum	4

More detailed work such as DNA base composition and DNA homology studies will be necessary to determine the correct systematic position of these isolates.
The collected data represent the first ecological approach to the occurrence of Azospirillum spp. in the Somalian soils, and confirm the high incidence of these microrganisms in the equatorial zone, as reported by other authors. (17-18).

References

1. Döbereiner J., Mariel J.E. and Nery M. (1976). Can. J.Microbiol. 22, 1464-1473
2. Tyler M.E., Milam J.R., Smith R.L., Schank S.C. and Zuberer D.A. (1979). Can.J.Microbiol. 25, 693-699
3. Haahtela K., Wartiovaara T., Sundman V. and Skujins J. (1981). Appl.Environ.Microbiol. 41, 203-206
4. Lamm R.R. and Neyra C.A. (1981). Can.J.Microbiol. 27, 1320-1325
5. Nioth J. (1983). Soil.Sci.Plant.Nutr., 29, 15-23
6. Favilli F., Balloni W. Cresta E. and Messini A. (1984). In Veeger, C. and Newton, W.E. (eds.). Advances in Nitrogen Fixation Research. Nijhoff, M. and Dr. Junk, W. publishers, The Hague. p.339
7. Florenzano G. and Favilli F. (1984). Giornale di Agricoltura, REDA 94, 40-41
8. Day J.M. and Döbereiner J. (1976). Soil Biol.Biochem. 8, 45-50
9. Rodriguez Caceres E.A. (1982). Appl.Environ.Microbiol. 44, 990-991
10. Tarrand J.J., Krieg N.R. and Döbereiner J. (1978). Can.J.Microbiol. 24, 947-980
11. Baldani V.L.D. and Döbereiner J. (1980). Soil Biol.Biochem.

12, 433-439

12. Döbereiner J. (1983).In Klingmüller,W. (ed.) Azospirillum II: Genetics, Physiology, Ecology. Experientia supplementum 48, Birkhäuser-Verlag, Basel, p.9-23
13. Hegazi N.A., Amer H.A. and Monib M. (1979). Soil Biol.Biochem. 11, 437-438
14. Markwell M.A.K., Haas S.M., Bieder L.L. and Tolbert N.E. (1978) . Anal.Biochem. 87, 206-210
15. Tomaselli L., Margheri M.C., Brancaccio V. and Florenzano W. (1984). In Veeger,C. and Newton, W.E. (eds.). Advances in Nitrogen Fixation Research. Nijhoff, M. and Dr. Junk, W. publishers, The Hague. p.362
16. Hylemon P.B., Wells J.S., Krieg N.R. and Jannash H.W. (1973). Intern.J.Sys.Bacteriol. 23, 340-380
17. Reynders L. and Vlassak K. (1982). In Klingmüller, W. (ed.) Azospirillum: Genetics, Physiology, Ecology. Experientia Supplementum 42, Birkhäuser-Verlag, Basel, p. 121-129
18. Rao A.V. and Venkateswarlu B. (1982). Can.J.Microbiol. 28, 778-782

TEMPERATURE AND SALT TOLERANCE OF *AZOSPIRILLUM* SPP. FROM SALT-AFFECTED SOILS IN BRAZIL

B. Reinhold[1], T. Hurek[1], I. Baldani[2], and J. Döbereiner[2]

[1]Institut für Biophysik, Universität Hannover, Herrenhäuser Str. 2, 3000 Hannover 21, FRG and [2]EMBRAPA-UAPNPBS, km 47, Seropedica, 23851 Rio de Janeiro, Brazil

Summary:Roots and rhizosphere soil of ten plant species from salt-affected soils in the arid northeast region of Brazil and from flooded, sea-water affected soils south of Rio de Janeiro were checked for presence of *Azospirillum halopraeferens* or bacteria of similar physiological properties. *A. halopraeferens* could not be detected in these samples. Among the 37 *Azospirillum* isolates, most were determined to be *A. brasilense*. Few strains belonged to *A. lipoferum* or had unusual characters. Two isolates were checked for DNA-DNA homology with type strains by the initial renaturation rate method. The biotin requiring strain 68 was found to be *A. brasilense*, strain 19 could not be assigned to any of the three species mentioned above. None of the isolates tested showed an optimum temperature of N_2-dependent growth at 41°C as *A. halopraeferens* does, but some isolates showed still significant acetylene reduction activity at 41°C. All strains tested were able to grow at this temperature on combined nitrogen, some of them with only slight decrease in doubling time. Unlike *A. halopraeferens*, none of the isolates tested showed better growth at increased NaCl concentrations (0.25%), but most of them sustained N_2-dependent growth at 0.75% NaCl better than strains Sp7 and Sp59b. Strain 68 even tolerated 3% NaCl for N_2-fixation. *Azospirillum* strains isolated from salt-affected soils in Brazil were more temperature and salt tolerant with respect to nitrogen fixation than *Azospirillum* species from other habitats and thus seem to show some adaptations to their environment in this respect. Keywords: *Azospirillum halopraeferens*, salt-affected soils, temperature tolerance, salt tolerance.

Azospirillum IV: Genetics, Physiology, Ecology
Edited by W. Klingmüller

Introduction

Recently, a fourth species of *Azospirillum*, *A. halopraeferens*, has been described (1). Strains had been isolated from the root surface of Kallar grass grown on salt-affected soils in the region of Punjab, Pakistan. Compared with all previously described *Azospirillum* species, they possess two unique properties: a high optimal temperature for growth (41°C) and a preference for higher NaCl concentrations (0.25%). This rose the question, if bacteria of this species or of similar physiological properties occur in the rhizosphere of other plants in other regions. As all other species of *Azospirillum* had been found to be widely distributed in Brazil (2, 3), Brazilian soils were chosen for the survey. In the arid northeast region of Brazil, areas of agricultural potential, are also salt-affected. Plants of this region and from flooded soils with assess to sea water south of Rio de Janeiro were used for this study.

Materials and Methods

Samples from the northeast were sent by air to Rio de Janeiro; samples from the coast near Rio were treated within 3h after collection.Rhizosphere soil, root pieces free of loosely adhering soil, and dilutions of homogenized washed roots were used to inoculate enrichment media. Enrichment was done in two semisolid synthetic malate media, SM medium (4) pH 6.5 without NH_4Cl and yeast extract, and the same medium adjusted to pH 7.2 and supplemented with 2.5 g/l of NaCl (DS medium). A mixture of vitamins (1) was added to both

Table 1 Typical reactions of type strains in tests used for identification of strains showing *Azospirillum* type morphology.

Test	*Azospirillum*			
	lipoferum	*brasilense*	*amazonense*	*halopraeferens*
Growth in N-free semisolid medium on				
glucose	+	-	+	-
mannitol	+	-	-	weak
Biotin requirement	+	-	-	+
Glucose acidification	+	-	-	-
Pleomorphism on alcaline media	+	-	-	-

of them. Bacteria were isolated by the pouring method (1) on the same media containing additionally 8 g/l of agar and 20 mg/l of yeast extract. Tests used for identification of the strains showing *Azospirillum* morphology were carried out as described previously (2, 5) at 35°C and are listed in Table 1. DNA-DNA homology was checked by the initial renaturation rate method (6). Salt and temperature tolerance were studied on semisolid N-free SM medium by using the acetylene reduction assay and on liquid SM medium by monitoring the optical density.

Results and Discussion

Occurrence of *Azospirillum* species in Brazilian rhizosphere samples. *Azospirillum* was found to be widely distributed in salt-affected soils. Out of eight different rhizosphere and root samples, we obtained 37 isolates showing *Azospirillum* type morphology (Table 2), several of them isolated at 41°C. No diazotrophs could be isolated at 41°C from diluted samples (MPN count vials) of homogenized roots of plants from the northeast, i.e. less than 10^4 diazotrophs per g root fresh weight were present; numbers might have been reduced during transportation. MPN counts of samples from the coastal region were positive for one grass when carried out at 32°C. Most of the strains were identified to be *Azospirillum brasilense*; few strains of *A. lipoferum* could be isolated at 32°C only. Several strains had unusual characters: strains 1s, 1sb, and 4r, differed from *A. brasilense* by their requirement of biotin and pleomorphism; strain 18 resembled *A. lipoferum* physiologically but *A. brasilense* morphologically; strain 68 and 19 differed from *A. halopraeferens* by their inability to use mannitol, and strain 19 additionally showed weak growth on glucose and pleomorphism. Unlike 1s, 1sb, and 4r, both strains had been isolated on DS medium and were therefore chosen for genetical comparison with *A. halopraeferens*. The average (G+C) content of the DNA determined by thermal denaturation was 70.8% for strain 68 and 71.3% for strain 19. No meaningful DNA-DNA binding (≤25%) was found between these isolates and *A. halopraeferens* Au 4. As a control, we found 96% DNA-DNA binding between strains Au 4 and Au 5 of the same species. Strain 68 could be assigned to *A. brasilense* by showing 69% binding with *A. brasilense* Sp7, but less than 25% binding with *A. lipoferum* Sp59b. Thus, biotin requirement also occurrs in *A. brasilense*. Strain 19 could not be assigned to one of these species as no

Table 2. Recovery of *Azospirillum* from roots or rhizosphere samples of the northeast of Brazil[a] and south of Rio de Janeiro[b]

Plant	Strain	Origin[c]	Isolation Temperature (°C)	Species affiliation[d]
Neoglaziovia variegata[a]	1s, 1sb	s	32	unusual (*lipoferum*)
Cordia leucocephala[a]	24	s	32	*brasilense*
Opuntia ficus-indica[a]	3, 4	s	41	*brasilense*
	19	s	32	unusual (*lipoferum*)
	20	s	32	*brasilense*
	3	r	32	*lipoferum*
Chloris gayana[a]	1, 5	r	41	*brasilense*
	6, 12, 16	s	41	*brasilense*
	18	r	32	*lipoferum*(*brasilense*)
	4r	r	32	unusual (*lipoferum*)
Anchrus ciliaris	7, 10, 14, 51, 59	s	41	*brasilense*
cv. *Biloela*[a]	56, 5r, 5r1, 5r2	r	32	*lipoferum*
Urochloa mosambicensis[a]	11, 15, 17	s	41	*brasilense*
	51	r	32	*brasilense*
unidentified grass[b]	60, 61, 62, 68	MPN	32	*brasilense*
Sedge[b]	64, 64, 67	r	32	*brasilense*
Paspalum sp.[b]	-			
Halicornia sp.[b]	-			

[a], [b] See heading.
[c] s, rhizosphere soil; r, roots; MPN, 10^3 fold dilutions of homogenized roots.
[d] If cell morphology not correlating with physiological tests, type of morphology indicated in parentheses; unusual, unusual physiological reactions.

meaningful DNA-DNA binding was found.

Azospirillum halopraeferens could not be detected in the Brazilian samples. The extreme environment where it is occurring might make a wide distribution more difficult. Bacteria resembling *Azospirillum* but having unusual characters have already been described by others (5); there might exist even more than four species.

Temperature and salt tolerance of isolates. As many strains had been isolated at 41°C and 2.5 g/l of NaCl, we were interested if these conditions were also optimal for *Azospirillum* spp. other than *A. halopraeferens*.

In contrast to *A. halopraeferens*, acetylene reduction activity of all isolates was not increased by the addition of 2.5 g/l of NaCl (Fig. 1, 2), but

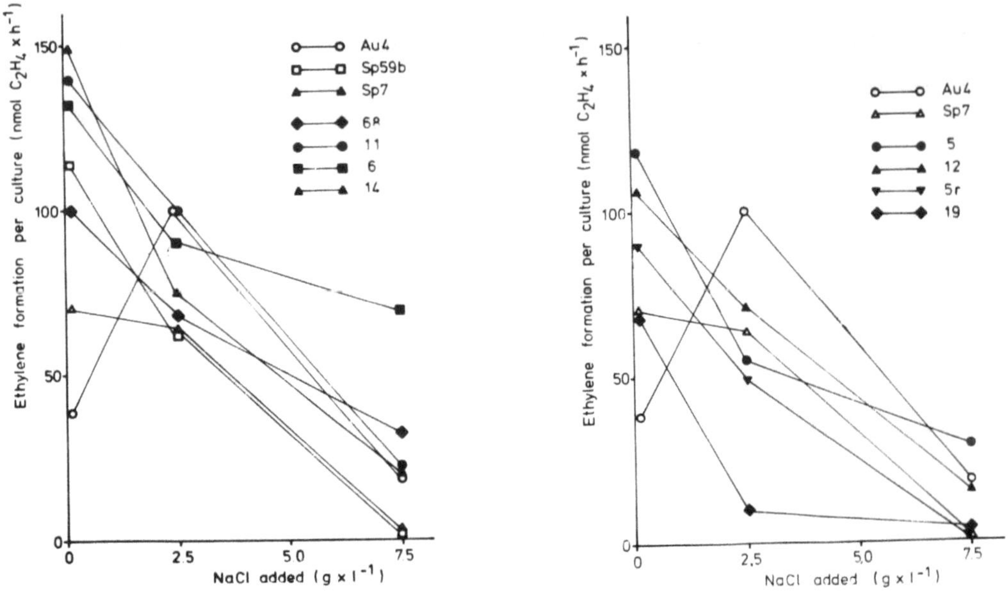

Figures 1 and 2 Effect of addition of NaCl on acetylene reduction activity. Inocula were precultured on liquid SM medium and washed twice in N-free medium. Strains were grown for 24h in N-free semisolid SM medium at 37°C, NaCl addition as indicated. After incubation with 10% acetylene for 1h, ethylene formation was measured gaschromatographically. Data are means of four replicates.

most isolates sustained N_2-dependent growth at 7.5 g/l of NaCl better than strains Sp7 and Sp59b. On 3% NaCl, besides *A. halopraeferens* only strain 68 developed an acetylene reducing pellicle within one week. When grown on

Table 3 Effect of addition of 0.25% NaCl to SM medium on doubling time (h) of *Azospirillum* spp. type strains and isolates

Strain	-NaCl	+NaCl	Strain	-NaCl	+NaCl
A. halopraeferens Au 4	8.9	3.1	56	1.7	1.9
A. brasilense Sp7	1.6	1.7	11	2.1	1.9
A. lipoferum Sp59b	2.1	2.5	20	1.6	2.2
68	2.2	2.1	60	2.2	2.1
19	1.9	2.0	12	1.8	1.8
6	1.3	1.4	5r	1.9	2.0
14	1.5	1.5	5	1.8	1.7

combined nitrogen, addition of 2.5 g/l of NaCl also decreased the doubling time dramatically in *A. halopraeferens* (Table 3). For the other strains, the effect of NaCl on growth was less pronounced with combined nitrogen than with N_2 as a nitrogen source. Doubling time decreased or increased only slightly. This

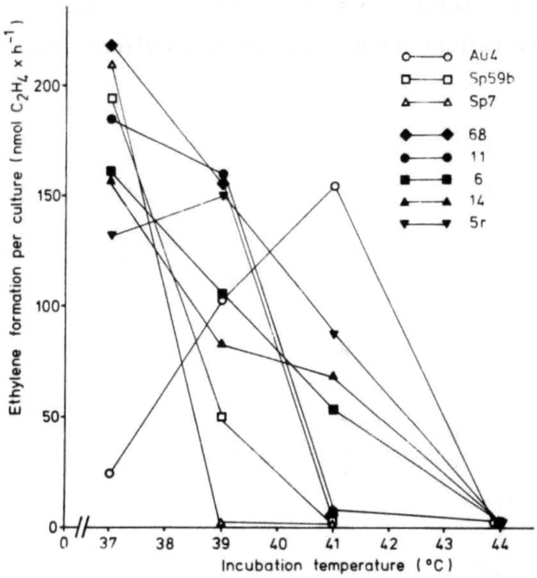

Figure 3 Effect of incubation temperature on acetylene reduction activity. Growth of precultures see Fig.1. Strains were grown for 24h at the temperature indicated on N-free semisolid SM medium (Au4 on DS medium). After incubation with 10% C_2H_2 for 1h, ethylene formation was measured gaschromatographically. Data are means of four replicates.

indicates that the nitrogen-fixing system is more sensitive to NaCl in other *Azospirillum* species as already proposed by Rao and Venkateswarlu (7). In *A. halopraeferens*, the pronounced effect which can be observed even on combined nitrogen is probably due to a sodium requirement (T. Hartmann, pers. comm.).

In contrast to *A. halopraeferens*, none of the isolates tested showed optimal N_2-dependent growth at 41°C, but some strains originating from the northeast were still able to grow on fixed nitrogen at this temperature in contrast to Sp7, Sp59b and isolate 68 originating from Rio de Janeiro (Fig.3). Isolates from *Digitaria* and maize had failed to grow on N_2 above 39°C (8). On combined N, all strains including Sp7 and Sp59b were able to grow at 41°C (Fig. 4), but failed to grow at 44°C. The relative temperature tolerance of the

N_2-fixing system especially of *A. halopraeferens* is remarkable and has not been shown in other N_2-fixing bacteria. N_2-fixation in cyanobacteria isolated from hot springs was observed even at 60°C (9). In *Klebsiella pneumonia*, the thermolability of N_2-fixation above 20°C was shown to be due to the *Nif*A type gene product (10).

Salt and temperature tolerance with respect to nitrogen fixation of Brazilian isolates probably reflect an adaptation to environmental conditions.

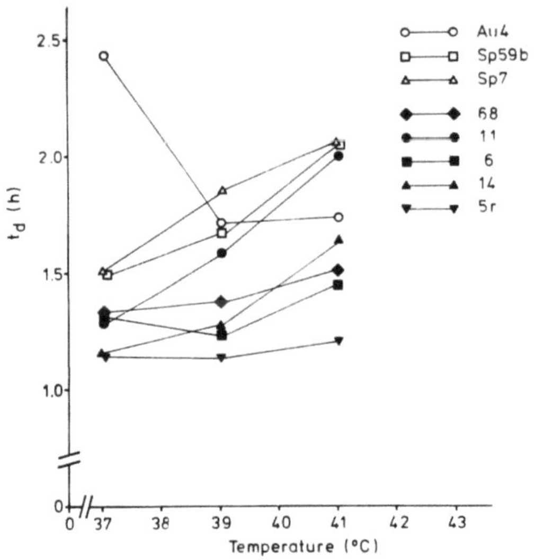

Figure 4 Effect of incubation temperature on doubling time. Growth of precultures see Fig. 1. Strains were grown on liquid SM medium (DS medium for Au 4), 30 ml of medium in 100ml Erlenmeyer flasks, with rotary shaking at 200 rpm. Doubling time was calculated from increase in optical density at 578nm. Data are means of two replicates.

Acknowledgements

This work was supported by Kernforschungszentrum Jülich under Brazilian-German cooperation. We thank Dr. G. Auling, Institut für Microbiology, Hannover, for making the Gilford 2600 spectrophotometer available to us for measurement of renaturation rates.

References

1. Reinhold, B., Hurek, T., Fendrik, I., Pot, B., Gillis, M., Kersters, K., Thielemans, S., and De Ley, J.(1987). Int. J. Syst. Bacteriol. 37, 43-51
2. Tarrand, J. J., Krieg, N. R., and Döbereiner, J.(1978). Can J. Microbiol. 24, 967-980
3. Magalhaes, F. M., Baldani, J. I., Souto, S. M., Kuykendall, J. R., and Döbereiner, J.(1983). An. Acad. Bras. Cienc. 55, 417-430
4. Reinhold, B., Hurek, T., and Fendrik, I.(1985). J. Bacteriol. 162, 190-195
5. Krieg, N. R., and Döbereiner, J.(1984). In Krieg, N. R., and Holt, H. G. (eds.):Bergey's manual of systematic bacteriology, The Williams and Wilkins Co., Baltimore, vol. 1, 94-104
6. De Ley, J., Cattoir, H., and Reynaerts, A.(1970). Eur. J. Biochem. 12,133-142
7. Rao, A. V., and Venkateswarlu, B.(1985). Acta Microbiol. Hung. 32, 221-224
8. Neves, M. C. P., Nery, M., and Day, J. M.(1976). An. XV Congr. Bras. Cienc. Sol. 143-149
9. Stewart, W. D. P. (1977). In Hardy, R. W. F., and Silver, W. S. (eds.): A treatise on dinitrogen fixation, Section III: Biology, John Wiley and Sons, New York, 63-123
10. Buchanan-Wollaston, V., Cannon, M. C., Beynon, J. L., and Cannon, F. C.(1981). Nature(London) 294, 776-778

AZOSPIRILLA FROM RICE FIELDS NEAR HANOI

Nguyen ngoc Dung, Nguyen phuong Chi and Ly kim Bang
Department for Microbial Technology,
Institute for Biology,
National Center for Scientific Research of Vietnam.

Summary: Azospirilla are frequent in the rhizosphere of rice plants in Tu-liem district. Isolates obtained were either typical Azospirilla or had at least some characters of Azospirilla. Members of both groups had beneficial effects on the roots of rice seedlings. Keywords: Isolation of new strains, rhizosphere, rice plants.

Introduction

Azospirilla are bacteria which have been studied extensively in recent years (1,2,3). Their wide distribution in the rhizosphere of certain plants belonging to the Poaceae in tropical and temperate regions is well documented. Our present work is directed towards determining the occurence of Azospirilla in the rhizosphere of rice plants near Hanoi.

Materials and Methods

Rhizosphere soils and roots of rice plants were used as samples. Two techniques for enrichment were used: One as described by Döbereiner and co-workers (see 4), the second was using the spermosphere system developed by Balandreau's groups (5). Samples are positive, if in semisolid Nfb-Medium after one or more days of growth a typical pellicle or veil is formed, the cells on microscopical inspection are vibroid with typical spiral movement, and C_2H_2 is reduced to C_2H_4 in the acetylene reduction assay.

Azospirillum IV: Genetics, Physiology, Ecology
Edited by W. Klingmüller

Results and discussion

With these methods, the occurence of Azospirilla in the rhizosphere and on the surface of rice roots in Tu-liem district, suburb of Hanoi, was widely detected (Table 1). To test for Azospirilla located inside the roots, roots were washed many times with steril water, then treated by a 10% H_2O_2-solution for 15 sec., washed with steril water again and crushed with a small steril glasrod. By this method, we could not identify Azospirilla inside roots.

Table 1: Occurence of Azospirilla in some places in Tu-liem

Place	pH of soil	Nr. of samples	Nr. of pos. samples	Isolate
garden of Nat. Center for Sc. Research of Viet.	6.5	3	3	Arl4, Arl14 Lun2-6-1
rice field in Nghia-do	6.5	3	3	Arm2-2,Arn4-1, Arm5-3
biol. Station in Co-nhue	6.0	5	5	Mat1-3,Mat1-4, Mat1-9,Mat2-1, Luc2-16-1

Azospirilla isolated by us were characterized as forming two groups. One of them contains strains Luc2-16-1, Lun2-6-1, Lun2-7-3, Mat1-3, Mat1-4, Mat1-9, Mat2-1 and Arm2-2. Their cells were typical vibroid and moved in spirals. They may be Azospirillum sensu strictu. Among them strain Arm2-2 was especially interesting for us. It was isolated by using the spermosphere system. On potato medium plates, on the first days, the colonies of Arm2-2 were flat and colourless. However after 10 days, they were slightly pink and their margin was slightly raised. The cells of Arm2-2 had a size of about 0.70 x 3.0 um and lateral flagella. According to Tarrand et al. (6) the strain Arm2-2 may be an Azospirillum which differs in some characteristics from the well known Azospirilla (Table 2).

Table 2: Differences between Arm2-2 and Azospirillum standard strains (according to Döbereiner, 7)

	A. amazonese	A. lipoferum	A. brasilense	Arm2-2
colonies on BMS	white, flat raised margin	pink raised	pink raised	pink, flat raised margin
cell width (um)	0.68+0.08	1.0 to 1.5	0.9+0.03	about 0.70
polar flagellum	+	+	+	+
lateral flagella on nutrient agar	-	+	+	+
biotin required	-	+	-	-
utilization of sucrose	+	-	-	+

The second group of isolates contained strains Arl4, Arl14, Arm4-1 and Arm5-3. They had some characters of Azospirillum, e.g. they formed a pellicle in the semisolid Nfb-Medium with malic acid or with sucrose as a C-source. They do not use glucose and can reduce C_2H_2 to C_2H_4. However they have a very small size (0.65 x 1.2 um) and their colonies are yellow.

Steril rice seedlings were inoculated with Arm2-2 and Arm5-3 at a concentration of 10^6 cells/ml. After 10 days the root growth of the seedlings was seen to be stimulated (Figure 1 and 2, "doi chung" means control).

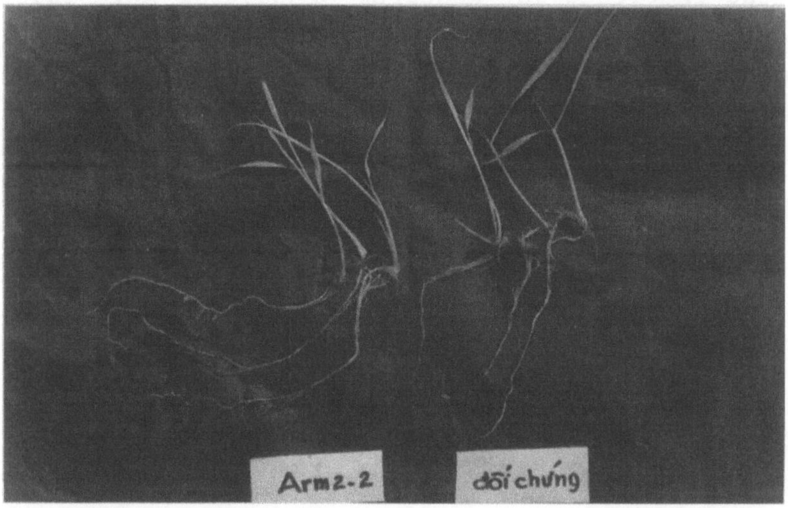

Fig. 1: Rice seedlings grown on minimal medium with 0.4 g KNO_3/l for 10 days. With and without Arm2-2 inoculation.

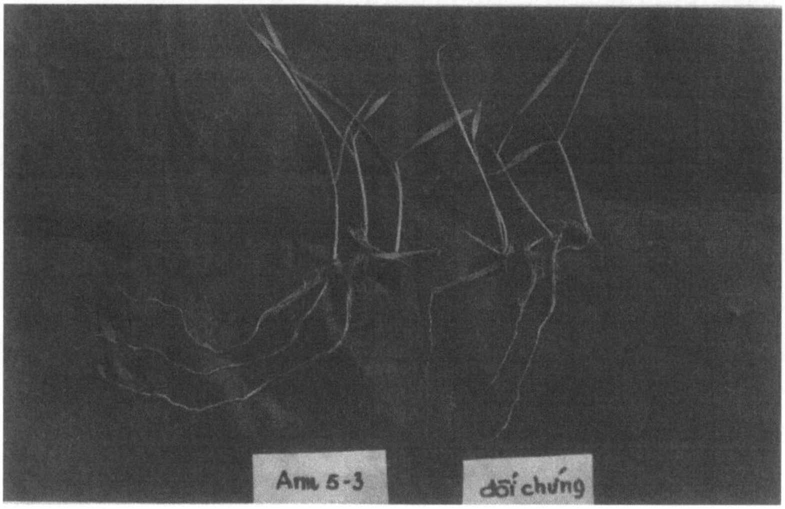

Fig. 2: Same as fig. 1, but with and without Arm5-3 inoculation.

Acknowledgement

We are grateful to Dr. Lai thi thuy Hien and the Department of Electrone Microscopy of the Institute for Army-Medicine, for the morphological fotographs and to Mrs. Nguyen thi Van, Department for Analyses of the Institute for Research of Natural Materials, for the gaschromatography. Also to Prof. Dr. W. Klingmüller, Genetics Department, Bayreuth University, FRG, for his effective help in all aspects.

References

1. Klingmüller, W. (ed.) (1982) Azospirillum: Genetics, Physiology, Ecology. Experientia Supplementum 42, Birkhäuser-Verlag, Basel

2. Klingmüller, W. (ed). (1983) Azospirillum II: Genetics, Physiology, Ecology. Experientia Supplementum 48, Birkhäuser-Verlag, Basel

3. Klingmüller, W. (ed.) (1985) Azospirillum III: Genetics, Physiology, Ecology. Springer-Verlag, Heidelberg

4. Becking, J.H. (1982) in quotation 1), p. 130-149

5. Heulin, T., Bally, R., and Balandreau, J. (1982). In quotation 1), p. 92-99

6. Tarrand, J.J., Krieg, N.R. and Döbereiner, J. (1984). In Krieg, N.R. and Holt, J.G. (eds.) Bergey's Manual of systematic Bacteriolgy, Vol. I. William & Wilkins, Baltimore

7. Döbereiner, J. (1983). In quotation 2), p. 9-23

MIX
Papier aus verantwortungsvollen Quellen
Paper from responsible sources
FSC® C105338

If you have any concerns about our products,
you can contact us on
ProductSafety@springernature.com

In case Publisher is established outside the EU,
the EU authorized representative is:
Springer Nature Customer Service Center GmbH
Europaplatz 3, 69115 Heidelberg, Germany

Printed by Libri Plureos GmbH
in Hamburg, Germany